Two-Stroke
TUNER'S HANDBOOK

By Gordon Jennings

Illustrations by the author

FOREWORD

Only ten years ago the two-stroke engine was widely and quite understandably thought to be a "reasonable alternative to the four-stroke only when minimum weight and manufacturing cost were all-important considerations. The two-stroke was recognized as having substantial theoretical promise, as it delivered a power stroke for each 360 degrees of crankshaft rotation but the hard reality was that each individual power impulse was too feeble to amount to much when totaled at the output end of the crankshaft. A very few engines had begun to appear in which some of the theoretical promise was realized however, and this encouraged engineers at MZ, Yamaha and Suzuki to persist in their efforts to wring competitive power output from the racing two-stroke engine. To say that they were ultimately successful would be gross understatement.

Those engineers were motivated by the need to demonstrate that the two-stroke engine, *per se,* was worthwhile -as that would stimulate sales of their companies' ordinary touring models. My own interest in the two-stroke, which had reached the level of an obsession by 1963, was generated by comparative poverty. I like to tinker with engines, and the complexities of the poppet-valve four-stroke make modifications very expensive. One may think that a change in valve timing would do wonders for a four-stroke's power, but getting a camshaft made to order costs hundreds of dollars. In contrast, a two-stroke engine's valve timing may be altered simply by reshaping the holes in its cylinders, and its power output markedly changed by utilizing inertia and resonant effects in its intake and exhaust tracts. None of these modifications are costly.

On the other hand, while the two-stroke engine does not commonly require large dollar inputs to raise its power output, it does require an in-depth under- standing on the part of the man doing the modifications. In an attempt to acquire that understanding I began a study of the high-speed, high-output two-stroke engine that has led to the collection of a minor library of text books and SAE papers. And to an endless series of experiments, some of them illuminating and many others raising more questions than they have answered. At this stage I have arrived at more or less satisfactory explanations for most of the gross phenomena, such as the general behavior of expansion chambers and port time-area values, and I flatter myself to think that just that much is an acceptable excuse for writing this book for the guidance of the layman experimenter. If it will not supply all of the answers it will at least take care of the fundamental problems and prevent the worst mistakes.

My special thanks to Mr. John Brooks, of McCulloch Engineering, who has done much to dilute my once pure ignorance (but should not be held accountable for the residue found herein). Also to the late Henry Koepke, who mistakenly assumed that I knew something about two-stroke engines and supported my early research; to my old friend Joe Parkhurst, who started me working on this book nearly ten years ago but never got it; and finally to Tom Heininger, who wheedled, needled, pleaded, complained and cajoled until I hammered my file of notes into publishable form.

CONTENTS

FUNDAMENTALS

Throughout this book it will be assumed, inconvenient though that assumption may occasionally be, that the reader has progressed to at least a superficial knowledge of the manner in which a piston-type internal combustion engine - with particular reference to those operating on the two-stroke cycle principle-converts quantities of fuel and air into useful power delivered at the end of its crankshaft. People who need enlightenment in that regard will find a wealth of explanatory literature collected on the shelves of any public library; no real purpose would be served by lingering over the matter here. Neither will I attempt to instruct you in the elementary mathematics and physics required to grasp much of what follows, as again the public library is an entirely adequate source of information. What will be provided is a kind of "state of the art" report about high-speed, high-output two-stroke engines for laymen-who in most cases do not have access to the literature (SAE papers, etc.) available to engineers and thus must rely upon hunches (often wrong) and folklore (almost invariably wrong) for guidance. Many have learned, to their sorrow, that it is distinctly possible to lavish enormous amounts of time and money on the two-stroke engine without realizing a return appropriate to the investment. The information to be provided here will not make you a Kaaden, or Naito; it will help you to avoid some of the more serious mistakes.

The first serious mistake a layman experimenter can make is to assume that those who designed and manufactured his particular engine didn't know what they were doing. In point of fact, the professional engineer knows very well, and if the engine in question is something other than what the experimenter has in mind, there are excellent reasons: all engines are compromised, from what you might consider an ideal, in the interest of manufacturing economy and broad usefulness. For example, ports may derive their shape as much from what the design engineer intended to be a low scrap-rate casting as from consideration of flow characteristics. In other words, even something like ports-design always will be influenced by the demands of mass-production manufacturing. Similarly, designing for mass-market sales implies that an engine must be agreeable to many different uses -even though that inevitably means that it will do no single thing particularly well. In these areas will we find the latitude for "improving" an engine, and one should always be mindful that the real task is simply to tailor a mass-use product to a very specific application- and that in the tailoring process one inevitably will incur all the various expenses the engine's designer has avoided. Hours of labor may be required to finish rough-cast ports; dollars will be spent correcting other things that are the creatures of manufacturing economies; power added at maximum revs will be power subtracted at lower crankshaft speeds, while the increased speeds required to obtain large improvements in power output will be paid for in terms of reliability.

Another mistake commonly made, sometimes even by those who have enjoyed some success in modifying two-stroke engines, is to believe in a kind of mechanistic magic. Bigger carburetors, higher compression ratios, altered port timings and expansion chambers often do bring an improvement in power output, but more and bigger is not magically, instantly better. All must work in concert with the basic engine, directed toward the particular application, before they constitute a genuine improvement. You cannot treat them as a voodoo incantation, hoping that if you mutter the right phrases and stir the chicken entrails in the prescribed manner, your mild-mannered, all-purpose chuffer will be transformed into a hyper-horsepower fire-breather. With a lot of luck, you might get that result; the chances heavily are that you won't.

With all the mysticism filtered out, horsepower at any given displacement is simply a function of average pressure in the cylinder during the power stroke and the rate at which power strokes occur, minus work absorbed by friction and scavenging. Raise pressure and/or the delivery rate of the power strokes, or reduce friction and pumping losses, and the engine's net output will rise. Unfortunately, there are limitations on all sides: Pressure must be limited because of thermal considerations (and is further limited by an engine's restricted ability to recharge its cylinder with a fresh air/fuel mixture between power strokes). The limit for power strokes per unit of time is established by what is tolerable in terms of crankshaft rotational speeds, and tolerable here is what the bearings, rod and piston will survive, in inertia loadings, for what you consider an acceptable service life; the design engineer has already expressed his opinion in this matter. Pumping losses can be reduced -relative to the mass flow through an engine - with a properly designed exhaust system, but otherwise are an inevitable and almost invariable consequence of pulling air from the atmosphere, moving it through the engine, and out the exhaust port. Some improvement in output may be obtained with reductions in friction, but the scope for such improvements is very small compared to what may be accomplished with cylinder pressure and engine speed.

Obviously, pressure in a cylinder will vary continuously throughout an engine's entire power stroke. Knowing what those pressures may be in a given engine is useful, but more useful still is knowing what they should and are likely to be, as such knowledge can keep you from that futile exercise commonly known as flogging a dead horse -and from believing a lot of lies about how much power various people are getting from their engines. Engineers have an overall efficiency rating called "brake mean effective pressure" (bmep), which they calculate by working their way back through torque readings observed on the dynamometer, the leverage provided by crankpin offset, and piston-crown area. Thus, bmep says little about peak cylinder pressures (those measurements being taken with a pressure transducer and oscilloscope) but it is an excellent relative indicator of performance and highly useful in projecting power output from a modified engine.

PREDICTING POWER

An average, well-developed stock engine intended for use in a sports / touring motorcycle will have a bmep of about 70 psi. It is possible, and I must stress that word "possible", to raise this to perhaps 115 psi- an improvement of some 64-percent, which (if accomplished) will yield a 64-percent increase in power output without raising the engine's operating speed. Similarly, a 64-percent increase in operating speed without a change in bmep would have the same effect on output. You will see this in the following formula for calculating horsepower:

$$BHP = \frac{PLAN}{33,000}$$

Where BHP is brake horsepower

P is brake mean effective pressure, in psi

L is piston stroke, in feet

A is the area of one piston, in square inches

N is the number of power strokes per minute

Obviously, when the values of L and A are held constant, as would be the case with an engine having a piston displacement at the limit established for a particular racing class, then increases in power may only be obtained by increasing the values for P and N -and you will find that in practice it is a lot easier to increase the latter than the former.

As already stated, bmep figures for stock, touring-type engines with flow-restricting air cleaners and mufflers, and with porting / carburetion compromised in favor of smooth low-speed running, will be around 70 psi. Typical figures for engines with porting and other plumbing arranged solely (and effectively) for maximum horsepower at peak revs would be about 115 psi - with a few small, highly-developed two-stroke engines operating up at 125 psi. The exact number will vary according to unit cylinder displacement and the width of an engine's useful power band, but one may reasonably expect that engines suitable for motocross will fall in the 85-95 psi range - with big cylinders tending toward the lower figure and small cylinders *vice versa.* Road racing engines, tuned to exert a maximum effort over a very narrow speed range, will usually show a bmep of 100-115 psi, and of course the same remarks regarding the influence of cylinder size apply.

These numbers have a usefulness beyond the mere satisfaction of vulgar curiosity: they may be used very profitably to determine an engine's suitability for some particular application. For example, they shed light on the future prospects of those who are trying to transform Kawasaki's F-5 "Bighorn" engine, a 350cc single, into a prime-mover capable of ending the Yamaha TD-2's absolute domination in road racing. Much has been made, by the Kawasaki's supporters, of the usefulness of a broader power range inherent with the F-5's disc-valve induction and the l00cc advantage it gets, over the TD-2, by having only a single cylinder (this, under the present American Motorcycle Association rules). Now while it is true that a racing motorcycle having a wide power band is easier for its rider to manage, and may offer an absolute if very slight advantage on short, extraordinarily twisty circuits, one must not overlook the fact that the TD-2 has been blessed with an excellent close-ratio transmission and a number of riders quite capable of coping with any problems introduced by the need for frequent gear changes. Viewed realistically, the situation facing any serious challenger to Yamaha's supremacy is one in which horsepower must be met with horsepower. And what are the Kawasaki's prospects of developing that kind of horsepower? Let's have a look at the numbers:

Assuming that the man who modifies the Kawasaki F-5 knows his business, but doesn't have all the development time in the world, (probability favors the latter far more than the former) then he very likely will arrive at a combination of porting, etc., good for a bmep of about 105 psi-which is about all that can be expected with a single cylinder of 350cc displacement. To expect more would be to ignore the considerable difficulties in scavenging efficiently the F-5's large-bore (3.17-inch) cylinder. Further assuming (and as we shall see later, this assumption is far from safe) that the F-5 engine will remain in one, working piece for the duration of a longish race with its rider observing a red-line of 9000 rpm, with a power peak at 8500 rpm, then,

$$\text{BHP} = \frac{105 \times 0.223 \times 7.789 \times 8500}{33{,}000}$$

$$\text{BHP} = 47.6$$

So, a well developed F-5 would deliver 47.6 brake horsepower. How does that compare with the Yamaha TD-2? With all the years that have gone into the TD-2's development, and giving due thought to Yamaha's proven expertise in these matters, it seems safe to assume that this engine would be operating with a bmep of 115 psi at its power peak- which seems to be at 11,000 rpm. Thus, working from those numbers and the 250cc Yamaha twin's bore/stroke dimensions of 56mm and 50mm, respectively,

$$\text{BHP} = \frac{115 \times 0.164 \times 3.81 \times 22{,}000}{33{,}000}$$

$$\text{BHP} = 48.0$$

Clearly then, those who would try to beat the Yamaha with a Kawasaki F-5 have taken upon themselves a task of considerable magnitude. The only bright spot in the picture, for them, is that while they are 0.4 bhp down on the Yamaha (assuming near-optimum work on their part) they probably will have the advantage in terms of average horsepower, figured from the moment a gear is engaged - when revs fall somewhat below those for peak horsepower -until the red-line is reached and it is time for a change to the next higher gear. There will be no advantage in frontal area, for although the F-5 engine is narrower than that of the TD-2, the fairing must be wide enough to shroud the rider, and the minimum width that requires is sufficient to encompass either engine. Moreover, moving from the theoretical to the practical for a moment, it is highly unlikely that the Kawasaki could be made as reliable at 8500 rpm as is the Yamaha at 11,000 rpm, and not because the F-5 engine is badly designed or shoddily constructed. The simple truth is that any single-cylinder 350cc engine with the F-5's bore / stroke dimensions and red-lined at 9000 rpm is going to be stressed very near its absolute limit - a limit imposed by the properties of available materials.

PISTON SPEED

All this asks the question, "How does one determine the limit, with regard to engine speed?" Unfortunately, establishing this limit with any precision is not only extremely difficult in terms of the mathematics involved, but also requires data concerning metallurgy, etc., seldom available outside the record-rooms of the factories from which the engines originate. Still, there are guide-lines which, if lacking in absolute precision, do at least have the virtue of simplicity, and will provide an indicator to keep us away from certain trouble. It is almost impossible to establish the point, in engine speed, between zero trouble and the possibility of trouble; there is much less difficulty in determining a red-line between *some* trouble and nothing *but* trouble.

A quick and easy method of establishing a limit for crankshaft speed is by working with piston speed. Actually, with "mean" piston speed: pistons do not travel at uniform velocity; they move from a dead stop at each end of their stroke, accelerate up to a maximum speed that often is in excess of 120 mph, and then brake to another complete stop. For convenience, we use just the mean piston speed and the safe limit for that, for engines having bore-stroke dimensions within the range considered normal for motorcycles, is about 4000 feet per minute. And mean piston speed may be calculated very easily by applying the following formula:

$C_m = 0.166 \times L \times N$

Where: C_m is mean piston speed, in feet per minute
L is stroke, in inches
N is crankshaft speed, in revolutions per minute

Thus, using again the Kawasaki F-5 engine as an example, with L being 2.68-inches and N given as 9000, we find that

$Cm = 0.166 \times 2.68 \times 9000$
$Cm = 4000$ ft/min

Here we have a theoretically-predicted limit that seems to agree quite closely with observable reality in the field: Reports from those actually racing modified F-5 Kawasaki's indicate that the engine does in fact retain acceptable (within the framework of that word's meaning in racing) reliability when red-lined at 9000 rpm, and ravels with horrifying abruptness if pressed further. Of course, it must be stressed here that few engines, the F-5 not excepted, retain more than marginal reliability at mean piston speeds of 4000 ft/min, and even this presupposes frequent replacement of the piston and the crank/rod bearings.

You will be on far more solid ground if your engine is not asked to endure mean piston speeds above 3500 ft/min. Anything above that takes an engine into the twilight zone of reliability, and the ground between 3500 ft/min and the near absolute limit of 4000 ft/min is covered with unpleasant possibilities, but these often may be minimized with the proper selection of materials and lubrication. I should note here that there are exceptions to this rule among some of the old-fashioned, long-stroke engines, which tend to have very light (and strong) reciprocating parts relative to their absolute stroke. An example that comes to mind is the Bultaco 125cc TSS, which had a stroke of no less 2.36-inches (decidedly long for a 125) but which would, in "factory" road racing trim run up to 11,500 rpm, just like the Yamaha TD-2 (with a much shorter, 1.97-inch stroke), and that represents a mean piston speed of 4500 rpm. Obviously, Bultaco held the opinion that the resulting thin-ish margin of reliability was acceptable, but their TSS never was as predictably trouble-free as Yamaha's TD-2, which at the same crankshaft speed (11,500) has a mean piston speed of only 3775 ft/min.

While on the subject of bore/stroke dimensions, I would like to say that there is much in favor of long stroke two-stroke cycle engines in many applications. They are not superior (as many people seem to think) compared to the present day short-stroke designs in terms of low-speed torque, as torque is entirely a function of displacement and bmep, and wholely unrelated to bore/stroke ratios. With a long stroke, there is (at any given displacement) a reduction in bore, and with it a loss of piston area against which gas pressure can exert its force, that exactly balances the loss of leverage in a short-stroke engine (which is, in turn, compensated by a gain in piston area). The only thing wrong with the long-stroke engine is that its crankshaft speed is limited by inertia loadings, and that in turn limits its absolute power potential as compared with the "modern" short-stroker. On the other hand, it is compensated by having a much more compact combustion chamber, which makes for more efficient burning, and by lower thermal loadings on the piston as a result of the smaller crown area into which heat from the combustion process may soak. Finally, there is an advantage in port area for the long-stroke design resulting from its relatively large cylinder wall area. This area increases in the long-stroke engine because displacement rises only in direct proportion to stroke, but is increased by a factor of 3.14 (the constant, pi) with enlargements in bore. These are very real advantages, but they are not enough, usually, to prevail against the short-stroke engine's sheer ability to rev. Crankshaft speed is the only thing subject to much juggling in the horsepower equation- and is a far more potent factor in determining power output than the relatively slight improvements in bmep obtainable with the marginally better combustion chamber and porting in the long-stroke engine. A 10-percent improvement in our Kawasaki F-5 engine's bmep (a large improvement indeed) would raise its output to 52.3 bhp; leave the bmep unchanged, but shorten the stroke and spin it 11,000 rpm and you would have 61.3 bhp. There is indeed no substitute for revs.

PISTON ACCELERATION

Sadly, while there is no substitute for revs, there are plenty of barriers: piston speed is one, as was already noted. But that is a rather indirect limit, as it ignores the fact that it is not speed so much as all the starting and stopping of pistons that does the damage, or at least the worst of any damage. The acceleration forces generated by the starting and stopping are felt even in an engine's main bearings, but they are at a peak in the connecting rod and piston and have a particularly disastrous effect on the latter, as any attempt to make a piston stronger is apt also to make it heavier-which aggravates the very situation the strengthening of the piston should improve. Even so, an engine's true Achilles heel, the problem that may most strongly resist solution, often is the disastrous effects piston acceleration may have on the piston's rings.

It often is thought, and quite wrongly, that rings maintain a seal between the piston and the cylinder's walls simply through their properties as springs. A little thought should convince you that such cannot be the case, for most rings, compressed in the process of installation, press outward against the cylinder with a force amounting to about 30 psi. Gas pressure in that cylinder may easily exceed 750 psi, and it should be obvious

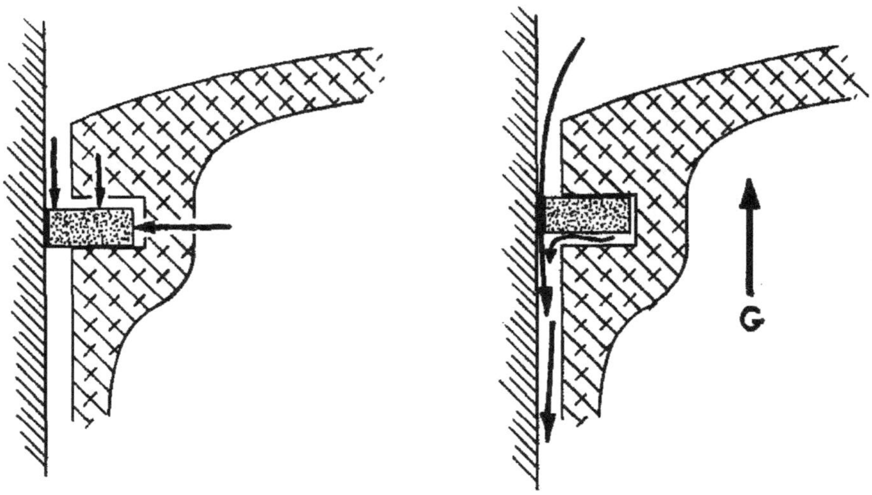

Normally gas pressure in the upper cylinder holds the ring down against the bottom of its groove and out against the cylinderwall, forming a seal (shown left). But piston acceleration, "G", can lift the ring, shut off pressure behind the ring and break the seal.

that a 30 psi force will not hold back one *circa* 750 psi. Still, equally obviously, piston rings do form an effective seal. How? Because they get a lot of help from the cylinder pressure itself: gas pressure above the ring forces it down against the bottom of its groove in the piston, and also (acting behind the ring, in the back of the groove) shoves it out hard against the cylinder wall. Thus, in the normal course of events, sealing pressure at the interface between cylinder wall and ring always is comfortably higher than the pressure it must hold back.

This very desirable situation will be maintained unless something happens to upset things, and most-insistent among the several "something's" that may intrude is excessive piston acceleration. When piston acceleration exceeds the sum total of gas pressures holding the ring in place, the ring will lift upward (as the piston nears the top of its stroke, and is being braked to a halt). Instantly, as the ring lifts, the gas pressure previously applied above and behind is also applied underneath the ring, at which point its inertia takes over completely and the ring slams up hard against the top of its groove. This last action releases all pressure from behind the ring, leaving it entirely to its own feeble devices in holding back the fire above, and as its 30 psi outward pressure is no match for the 750 psi pressure in the upper cylinder, it is blown violently back into its groove. The ring's radial collapse opens a direct path down the cylinder wall for the high temperature and pressure combustion gases-but only for a microsecond, for the action just described instantly applies gas pressure once again behind the ring and that sends its snapping back into place against the cylinder wall. Unhappily, it cannot remain there, as gas pressure immediately bangs it back into its groove again- to repeat the process over and over until the piston is virtually stopped and the ring's inertia is no longer enough to counter gas pressure.

The net result of all this activity is that over the span of several degrees of crank rotation, immediately preceding the piston's reaching top center, the ring will be repeatedly collapsed radially and at the same time hammered hard against the top of its groove. Understandably, the ring is distressed by this, as it not only receives a fearful battering but also is bathed in fire while being deprived of the close contact with piston and cylinder that would otherwise serve to draw off heat. Equally damaging is that the piston is having much the same problem, with high-temperature gases blowing down past its skirt to cause overheating, to burn away the film of oil between itself and the cylinder wall, and with its ring, or rings, all the while trying to pound their way up through the piston crown. A mild case of what is quite accurately termed "ring flutter" eventually results in the destruction of the ring and sometimes the dimensional integrity of its groove; a more serious case is certain to lead rapidly into lubrication failure, overheating, and piston seizure. Fortunately, this drastic problem can be avoided, thanks to the work of the researcher Paul de K. Dykes, whose investigation of the ring flutter phenomenon yielded most of what we know about it - and who invented the flutter-resistant ring that bears his name. Dykes showed us the cause of ring flutter, and engineers' understanding of the cause is reflected in their designs of the modern piston ring, which is very thin, axially, with a very considerable width, radially. Thus, gas pressure bears down on a large surface, providing an equally large total down-force, but is opposed by a relatively small upward load as the ring, being thin, is light and in consequence has little inertia. Still, even with very thin rings, flutter will occur if inertia loadings are high enough. To settle the question, with regard to any given engine, apply the following formula for determining maximum piston acceleration:

$$G_{max} = \frac{N^2 \times L}{2189}\left(1 + \frac{1}{2A}\right)$$

Where G_{max} is maximum piston acceleration, in feet per second squared
N is crankshaft speed, in revolutions per minute
L is stroke, in inches
A is the ratio of connecting rod length, between centers, to stroke

To illustrate how high these forces may sometimes be, let's use as an example the Yamaha TD-2, using 11,000 rpm for N. The formula tells us that at that speed, maximum piston acceleration will be (with the answer rounded off by my slide rule; I'm too lazy to do it all with paper and pencil) no less than 135,000 ft/sec2. Now if you will recall for a moment that the acceleration of gravity is only 32 ft/sec2, it will be clear that the load on the Yamaha's pistons - and thus on its rings - is very high indeed. But is the loading high enough to make the Yamaha's rings flutter? Obviously, it is not, as the engine remains not only reliable but crisp in comparatively long races. The limit, for the TD-2 engine, is slightly higher than 135,000 ft/sec2 - but not much higher, as you will see in the following table listing ring thicknesses and the accelerations at which they begin to flutter.

For rings having a	0.125-inch thickness,			40,000 ft/sec^2
	0.094	"	"	53,000 ft/sec^2
	0.063	"	"	80,000 ft/sec^2
	0.047	"	"	106,000 ft/sec^2
	0.039	"	"	138,000 ft/sec^2

The Yamaha, with rings having a thickness of 1mm, or 0.039-inch, and a maximum piston acceleration of 135,000 ft/sec2 at 11,000 rpm, would seem to be operating very near the limit - as indeed it is. But it probably is not quite as near the limit as the numbers suggest, for a racing ring (with its exaggerated thickness/width cross-sectional aspect) is somewhat less subject to flutter than a ring made for application in a touring engine. Still, the numbers given are fairly close for rings with normal-range proportions, and if you have an engine with rings for which Butter is predicted at 80,000 ft/sec^2 and intend using crankshaft speeds that would raise maximum piston acceleration to something more like 100,000 ft/sec^2, then I strongly urge you to fit new pistons with thinner rings. You may interpolate between the figures given to find the safe acceleration levels for ring thicknesses not listed.

There are piston rings that resist very strongly piston acceleration's efforts toward making them flutter. The best known of these is the Dykes-pattern ring, which has an L-shaped cross-section and fits into a similarly-shaped groove in the piston. The Dykes ring is made flutter-resistant by the fact that its horizontal leg fits quite closely in its groove, as compared to clearances around the vertical leg, and therefore even if acceleration lifts the ring it cannot lift high enough to close off the pressure behind the ring's vertical leg. In consequence, the ring's sealing abilities are maintained at accelerations that would be the undoing of rings in the conventional rectangular-section pattern. However, the Dykes ring's ability to maintain a seal does not free it of all the unpleasantness attending too-high piston acceleration: while it may seal under those

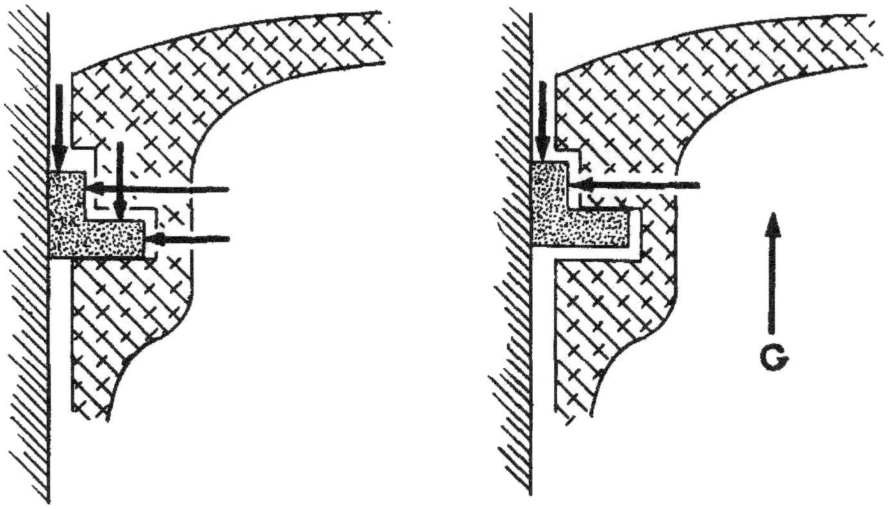

Dykes-pattern rings are subject to the effects of piston acceleration, "G", but gas pressure is maintained behind the ring because its horizontal leg limits its vertical travel.

conditions, it is still being rattled about vigorously and if the rattling continues long enough, the Dykes ring, and the groove trying to restrain it, both become badly battered. At that point, its ability to seal vanishes and mechanical failure of the ring, piston, or both, follows very closely. Bultaco has long used Dykes-pattern rings, as have certain others, but most manufacturers prefer rings that do not require such careful and intricate machining. There are other flutter-resistant rings, and many excellent reasons for using rings of conventional configuration, but these details are discussed elsewhere in this book and in greater depth than would be appropriate here.

After establishing all these mechanical limits, with regard to piston speed and acceleration, and after deciding how much power you are likely to get from a particular engine, you should subject the engine to a complete survey. This would include the measuring of port heights and widths, combustion chamber and crankcase volumes, and charting piston travel against crank rotation. This last effort may at first seem rather pointless, but as your work progresses you will find that the chart, which will show almost but not quite a sine curve, provides an instant readout between degrees at the crankshaft and the position of the piston from top center that is most useful. It will tell you, for example, how much to raise the top edge of an exhaust port to make a given change in timing, and how much to trim from the piston skirt (in a piston-port engine) to get the intake period you want-or think you want. The chart also will provide you with all the mean port-open points, and it will provide an exceedingly useful relationship between ignition timing expressed in degrees and in piston travel from top center. You may devise your own methods for deriving all this information according to your preference and resources; I have explained my own techniques elsewhere in this text, in the appropriate chapters.

An item that must be included in any discussion of the two-stroke cycle engine's basics is general gas dynamics. You can get information on the subject at your local library, but the applicable particulars are likely to be widely scattered there, so I will cover the subject in brief here. The manner in which what follows applies at specific points throughout the engine and its related plumbing will be covered later, but you should know a few of the fundamentals now and thus save me from becoming unnecessarily repetitious later.

One thing you must know, for example, is that the air moving through the engine, a mixture of gases, has many of the properties of a fluid. It even has the ability to "wet" a surface, and has viscosity, which means that air will cling to all surfaces within an engine in a layer that moves hardly at all no matter what the midstream velocity may be. This boundary layer's depth is influenced by gas temperature, and by the temperature of the surface on which it forms, as well as by the shape of the surface. Please understand that the layer is not solid; it is "shearing" with general flow throughout its depth – which may be as much as 0.100-inch - with movement increasing as to distance from the surface on which it is formed. And as close as 0.020-inch from the surface, flow may still be in the order of 80-percent of that in midstream, which means that the restriction formed by the boundary layer is not very great. Nonetheless, it is there, and it accounts for such things as round ports having less resistance to flow than square ports, area for area, and for the ability of a single port to match the flow of a pair of ports of somewhat larger area. It also accounts for the fact that flow resistance increases in direct proportion with the length of a port, and much of the resistance resulting from the shape of a particular port is due to that shape's creating a thick boundary layer, which becomes literally a plug inside the port.

Generally speaking, boundary layers will be held to minimum depth on surfaces that "rise" (relative to the direction of flow) and gain in thickness on any surface that falls away. Thus, an intake trumpet, for example, should be tapered in slightly from the inlet end to the carburetor-by perhaps 2-3 degrees - in the interest of holding boundary layer thickness to a minimum. In that configuration, it will have appreciably less resistance to flow than a straight, parallel-wall tube. Similarly, transfer ports should diminish in cross-sectional area from their entrance in the crankcase toward their outlet in the cylinder.

These gases also have inertia: once set in motion they tend to remain in motion; when at rest they resist all efforts to get them moving. In practice, this means that there always is a lag between the intake port's opening and the movement of air in the intake tract. Fortunately, this lag can be amply compensated toward the end of the intake period, when the pressure inside the crankcase has risen to a level that should push part of the charge back out the port-but cannot because of the effect of inertia on the incoming gases. Inertia also has its effect on the flow of gases through the transfer ports and out the exhaust system, but I will deal with that while treating those subjects separately.

These inertia effects are useful, but difficult to manage as something apart from other processes occurring as the engine runs. For example, intake tract length usually is established more with an eye toward resonances than inertia, and its diameter set by the flow rate required by the carburetor to meter properly - balanced against the resistance that attends high gas velocities. Therefore, virtually the only thing we can do about inertia effects is to attempt to find the intake timing that will make maximum use of those provided by an intake system proportioned mostly to suit other requirements.

Resonances are another matter. Sound waves will travel through any elastic medium, such as air, and in their passage they pull together or force apart molecules, just as the similar energy waves traveling through the ocean pull the water into peaks and troughs on its surface. And, as in the ocean, the waves move steadily onward away from their source but the transmitting medium does not. Take, for example, the activity surrounding a single condensation, or positive-pressure wave, as it moves through the air. In its center, molecules have been pulled together, condensed, but as it travels it releases those molecules and compresses others as it reaches them. In the same manner, a rarefaction, or negative-pressure wave, pushes molecules apart. Both waves behave in a curious, but useful way when confined in a tube and the effects of inertia are mixed with them. For one thing, they will be reflected back when reaching the end of the tube - whether that end is open or closed. But at the tube's open end, the wave changes in sign: a condensation is inverted and becomes a rarefaction, and *vice versa;* at the closed end, the wave will be reflected, but retains its sign.

How is all that useful? For example, in the intake system the opening of the intake port exposes the crankcase end of the tract to a partial vacuum, and that in turn sends a rarefaction shooting off toward the opposite, atmospheric, end of the tract. It travels out to the intake bell, inverts in sign to become a condensation, and instantly moves back toward the crankcase - to arrive there as a clump of compressed molecules, which surge into the crankcase to be trapped, if the piston then closes the intake port, as part of the scavenging charge. That effect, over-layed with inertia in the inrushing gases, makes all the difference in getting the job of charging done in two-stroke engines – which provide only an absurdly short time for such chores.

How short a time? That is at the same time one of the least complicated and most depressing calculations you can perform. Let us consider the Yamaha DT-1, which in fully developed configuration had an intake duration of 160-degrees, a transfer duration of 123-degrees, and an exhaust duration of 172-degrees. Yamaha claims a power peak at 7000 rpm. Let's have a look at the actual time, in fractions of a second, available for the completion of these functions. To arrive at these times, use the following formula:

$$T = \frac{60}{N} \times \frac{\theta}{360}$$

Where T is time, in seconds
N is crankshaft speed, in revolutions per minute
θ is port open duration, in degrees
(This formula can be abbreviated to $T = \dfrac{\theta}{N \times 6}$)

Thus, to find T for the 160-degree intake duration,

$$T = \frac{60}{7000} \times \frac{160}{360} = 0.0038 - \text{sec}.$$

With application of the same formula to the transfer and exhaust periods, we find that the former is open 0.0029-second, and the latter open 0.0041-second. Even the longest of these, the exhaust-open duration, is only 41/10,000-second, and that is not very much time in which to empty exhaust gases out of the cylinder. Actually, that particular process is substantially finished in the 29-degrees, or 0.0007-second, between exhaust- and transfer-opening. In that short period, pressure in the cylinder must fall to something very near atmospheric, or the exhaust gases would force their way down into the crankcase through the transfer ports. Of course, the exhaust gases are provided quite a large aperture by means of which they may make their escape, and that they do so, successfully, is less remarkable than the fact that the fresh charge compressed in a two-stroke engine's crankcase is able to make its way through the far more restricted transfer ports, propelled by a far lower pressure, to refill the cylinder in the extremely brief moment available. It seems nothing short of astonishing that this recharging operation is accomplished in the 0.0027-sec provided by the Yamaha DT-1's 114-degree transfer period; that the same process takes place in a Yamaha TD-2 engine in only 0.0017-sec appears a minor miracle. Obviously, divine intervention is not really a factor in the functioning of two-stroke engines, and cylinder recharging is possible simply because the process gets a lot of help from the activities of the exhaust system, gas velocities through the transfer ports have a mean value in the order of 300 ft/sec, and the cross-sectional areas of the ports involved are relatively large as compared with the volume of gases to be transferred.

As it happens, it is possible to calculate correct combinations of port-open times and port areas for any motorcycle engine, at any engine speed. The maximum safe speed for any engine is also calculable, as explained earlier in this chapter, along with expansion chamber dimensions, carburetor size and many other factors influencing both maximum power output and overall power characteristics. It should be noted here that none of the values derived purely from calculations are necessarily optima, and fine adjustments must always be made experimentally, but it is far better to employ the simple formulae presented in the chapters to follow than to attempt a purely-experimental approach. The mathematics involved are not terribly complicated, though sometimes the arithmetic is laborious, and you can use paper and pencil to arrive at a basic engine/pipe combination that will be very near the optimum. Much nearer, in fact, than would be obtained by even the most experienced tuner's unsupported guesswork, and near enough to a fully developed configuration to minimize the outlay of time and money entailed in the building of a racing engine. You start by determining, mathematically, an upper limit for engine speed, then use more math in establishing a maximum for piston-ring thickness, in establishing all the port dimensions to suit the projected engine speed, in selecting a carburetor, and in designing an expansion chamber. Suitable values for compression ratios, both primary and secondary, are provided in the chapters dealing with crankcase pumping and cylinder heads, respectively, and with the rest of the material included in this book it all adds up to being a fairly complete engine redesign manual for the two-stroke engine-fixated “tuner”. My own experience indicates that engines built along the lines suggested here never fail to deliver high specific horsepower (which is more than may be said for any cut-and-try system) even without the benefit of experiment-indicated adjustments. I dislike guesswork, have made a serious effort to eliminate it from my own projects, and am hopeful that the lessons learned - and outlined in this text - will reduce the generally high level of guesswork among most experimenters. If I have forgotten to cover anything, the omission is inadvertent, because my distaste for Speed Secrets is even greater than for guesswork. There is only one “Secret” in the game: to know what you are doing, and to do it thoroughly.

THE CRANK TRAIN

As was noted in the chapter of this book dealing with basics, power output from an engine of any given displacement is a function of gas pressure in the cylinder during the power stroke, and the number of power strokes per unit time. Implicit therein is the suggestion that the horsepower ultimately to be had from an engine has little to do with port shapes and port timings, exhaust systems, carburetion or indeed any of the things on which our attention usually is fixed. Why? For one thing, increases in gas pressure bring corresponding increases in heat flow into the piston -and no high-output two-stroke engine can operate beyond its thermal limit. Similarly, you cannot increase the rate at which power strokes occur without increasing crankshaft speeds, with increases in this direction sooner or later taking you beyond the engine's mechanical limit. The horsepower you ultimately will extract from any given engine depends therefore very directly upon your ability to expand those thermal and mechanical limits to the greatest extent possible, and only then to make the most of the territory thus gained.

THE PISTON

For a very long time subsequent to Dugald Clerk's creation of the two-stroke engine, the thermal limit was the only limit, but it was enough to hold power output from such engines to extremely modest levels. Then, as now, it was primarily a limit imposed by available piston materials. Cast-iron has its advantages in terms of wear resistance, hot-strength and low thermal expansion rates, and it was used quite frequently in the low speed engines of years past. Unfortunately, iron is heavy, and heavy is the last thing you want in a piston - which in modern engines is subjected to accelerations well in excess of 100,000 ft/sec^2. Aluminum, used as the primary constituent in virtually all piston alloys today, is conveniently light, but disagreeably insists on melting at much lower temperatures than that of the fire to which it is directly exposed. Moreover, it loses strength very rapidly with increases in temperature above ambient, so that piston failures do occur at crown temperatures well below the material's melting point. Finally, aluminum is a high expansion-rate metal, which makes a piston made of it a variable-clearance fit in any cylinder. But aluminum is a very light metal, and that alone was enough to recommend it for use in pistons, even though the drawbacks listed were enough to severely limit the specific power outputs attainable with two-stroke engines for a long time.

Aluminum-based piston alloys improved slowly over the years, with the addition of small percentages of, say, copper, to improve their hot-strength, but it was not until means were found to add considerable amounts of silicon that large improvements were made. Today, the best piston alloys contain between 15- and 25-percent silicon, and this addition has all but transformed the “aluminum” piston. Admixtures of silicon in excess of 15-percent not only drastically reduce aluminum's expansion rate, they also affect a proportionate increase in hot-strength and improve the piston's wear-resistant properties. In all of these respects the improvement is large enough to almost exactly equal the percentage gains in horsepower during the years in which aluminum-silicon alloys have been in use. I am inclined to think that most of what we consider to be "modern" improvements in two-stroke engine design – with particular reference to expansion-chamber type exhaust systems -might have been applied as much as fifty years ago had good pistons been available. There was little point in such development work without the aluminum-silicon piston; aluminum or aluminum-copper pistons would melt at specific power outputs well below what we now consider only average.

With all that, high silicon-content piston alloys still are not universally employed. As it happens, such alloys do have their disadvantage, which is that they are difficult to manufacture. Just casting pistons of aluminum-silicon alloy is a task for specialists using specialized equipment; machining the raw castings into finished pistons is an even more formidable task. You may encounter this last difficulty if you have occasion to modify a cylinder cast from the material in question - and you will find that it blunts cutting tools of any kind with remarkable rapidity. For you, that will be an inconvenience; for the mass-producer of pistons it is a disaster, as the need for frequent re-sharpening of tool bits entails losing output from his machinery while such repairs are made, and it means the expense of the man-hours required for the repairs. Thus, the manufacturer has every reason to restrict the silicon content of the piston alloys he uses to the minimum required by the use to which his engines will be put, which is the reason why Yamaha, for example, uses different alloys for touring and racing pistons.

In point of fact, the Japanese seem to manage high silicon-content pistons better than anyone else, which may well account for their notable superiority in coaxing power from two-stroke motorcycle engines. All of the major Japanese manufacturers employ piston alloys in their touring engines having percentages of silicon high enough to be considered "racing only" in much of the rest of the world. And, sad to say, many of the "racing" pistons being offered by speed equipment manufacturers are inferior in this regard to the ordinary off-the-shelf parts you'll find at your local dealer in Japanese motorcycles. For that reason, I am inclined to use either stock or "GYT-kit" pistons when I am working with engines carrying a "made in Japan" label, rather than waste my money on a specialty replacement. There are, of course, exceptions to this rule, which evolve principally around ring widths, and I will deal with that in due course.

Unless you happen to be a piston manufacturer, there isn't much you can do about piston alloys, beyond seeking out pistons having a high silicon content. Neither is there anything you can do about piston shape - which is most unfortunate, because a piston is

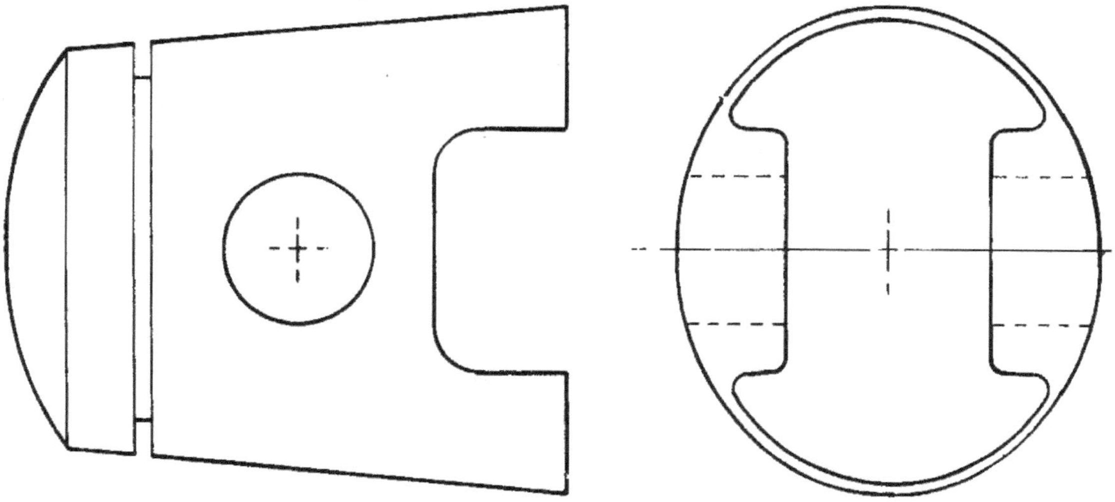

The cam and taper of the piston shown here are exaggerated to illustrate a piston's shape at ambient temperature. When hot, expansion changes the piston to fit the engine's cylindrical bore.

not, as it first appears, simply cylindrical. Even with the use of aluminum-silicon alloys, pistons do expand as they are heated, and they do not expand at all evenly. The greatest increase in diameter will occur up at the crown, because that is both the area of maximum mass and highest temperature. So there must be more clearance, measured cold, up at the piston's crown than is required down around the lower skirt. In fact, clearances vary continuously from the piston's crown to the bottom of its skirt -and from side to side, as the piston is elliptical rather than round. Someday, someone may be able, with the help of a computer, to actually calculate all the clearances and ellipse ratios involved; for the present they are decided in a process of trial-and-error by even the most experienced of manufacturers.

Presumably, you will not have the facilities to alter whatever shape your engine's piston(s) may have, but you can vary running clearances by changing cylinder bore diameter. The problem here is one of "How much?" and I regret to say that it is a problem for which there is no convenient solution. Clearances, measured at the piston's maximum diameter, across its thrust faces, may vary from about 0.002 to as much as 0.007-inch, depending on: the shape and composition of the piston itself; the absolute cylinder bore diameter; the material from which the cylinder is made, as well as its configuration; and the thermal loadings to which the piston will be subjected - which will themselves vary according to gas pressure, fuel mixture, cylinder configuration and the vehicle's rate of motion. Many people have expressed great faith in rules relating clearance to cylinder bore diameter; I have not found the choice to be that simple. If there is a rule, it would be that you can add perhaps 0.0005- to 0.001-inch to the clearance recommended by your engine's maker, but even this is a gross over-simplification and I mention it only because it is somewhat better to have too much clearance than too little. In the former, the excessive clearance adversely influences heat transfer from the piston to the relatively cooler cylinder walls and may lead to any of the several unpleasantries associated with overheating the piston, which range from a tendency for oil to become carbonized in the ring grooves, to the appearance of a large hole in the piston crown. Too little clearance will reveal itself in the form of scuffing, or outright seizure - unless the piston is only marginally too tight, in which case the only symptom of distress will be a power loss in the order of 2- to 3-percent.

Often, in modified engines, you will find that the straightforward increase in overall piston clearance by slightly enlarging the cylinder bore is not a complete answer. If the manufacturer has done his work properly, his pistons will, as they expand with temperature, assume a round shape when the engine is hot. Your problem will be that with the modifications you have made, more heat will be forced into the piston's crown, raising its temperature above the level anticipated by the manufacturer, which results in a completely different set of temperature gradients down the length of the piston. Specifically, while the whole piston will assume a diameter slightly larger than that planned for by its maker, the area around the crown will "grow" more than the rest. It will thus be impossible to correct for the altered conditions simply by honing the cylinder bore larger, for if you enlarge the bore enough to provide running clearance for the top of the piston, its skirt will be given too much clearance (leading to rocking, and trouble with the rings). In such cases, which are not the exception, but the rule, the solution is to machine what is called a "clearance band" around the top of the piston. Usually, this band will extend down from the crown to a point about 0.125-inch below the ring groove, or grooves, and the piston's diameter reduced by perhaps 0.002-inch over the entire band's width. Although the clearance band is not a particularly clean solution to the piston-expansion problem, it is one that can be applied by anyone with access to a lathe, and it has one advantage over the generally more desirable "pure" contouring of the piston: if a piston with a clearance band seizes partially, aluminum will not be smeared above and below the ring groove - an event which will lock the ring in its groove and upset its ability to seal against gas pressure, In practical terms, this means that the

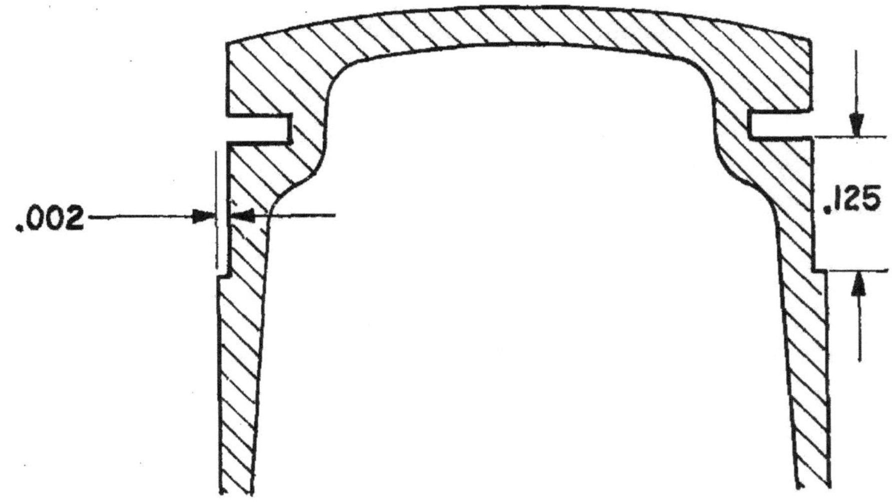

Stock pistons are not contoured to cope with the expansion produced by temperatures in modified engines and if frequently is necessary to machine a clearance band around the piston as illustrated here.

clearance-banded piston will absorb a lot of punishment before it is damaged sufficiently to cause retirement from a race.

Excessive deep clearance bands must be avoided, for they expose the sealing ring to too much heat, and heat has a devastating effect on the service life of a piston ring. But for these effects, there would be every reason to locate the ring as close to the piston crown as is mechanically possible, because we would then obtain the cleanest opening and closing of the ports; with the ring in its usual position, about 0.200-inch below the piston crown, there is a tendency for gases to leak down the side of the piston, and the port-opening process thus becomes more gradual than is desirable. The effect is slight, but it is there, and for that reason ring location always is a matter of juggling the conflicting requirements of keeping the ring cool, and obtaining sharp, clean port-opening characteristics. And in most instances, the balance of this compromise will be in favor of the former, for an overheated ring quickly fails. The cause of this failure is twofold: first, excessively high temperatures effectively anneal the ring, and it loses its radial tension; second, an overheated ring warps like a potato chip, and no longer maintains close contact with the bottom of its groove. In both of these cases, the ring's ability to seal is reduced, which allows fire to start leaking down past the ring, and that further raises its temperature -starting a cycle that soon results in outright ring failure.

The single exception to the unpleasantness just described is the L-shaped "Dykes" ring, which also is excepted from the immediate effects of ring-flutter (described elsewhere). A number of engines have been fitted very successfully with Dykes rings located right at the tops of their pistons, and the dire effects of excessive heating are avoided because the Dykes ring's vertical leg has enough area in contact with the cooler cylinder wall to draw away heat faster than it can be added by the ring's contact with high-temperature gases. At least, that's the way the situation can be, if everything is right. On the other hand, it is worth remembering that many users of Dykes-pattern rings have been forced to fabricate them from stainless alloys to overcome temperature related troubles, and even then have experienced problems with oil carbonizing in the ring grooves. Probably the best thing to be said for Dykes-pattern rings from the experimenter's viewpoint is that they can be used to overcome the problem of using stock pistons at very much higher than stock crankshaft speeds. If, for example, you would like to use the stock piston, but cannot because it has been grooved for rings 2.0mm thick and you must use 1.5mm rings to avoid ring flutter, you can simply cut a new groove at the top of the piston for a Dykes ring and the problem is solved -unless you encounter some of the other difficulties just discussed.

PISTON RINGS

Of all the problems that can be experienced with a modified engine, those connected with the pistons' rings are the most insidious. Borderline sealing failures can send fire shooting down along the pistons' sides to cause seizures and/or holing of the piston crown that appear to be the result of lean mixture, excessive ignition advance or too-high compression, but are not. These failures are, I suspect, much more frequent than is commonly supposed, for the 2.0mm rings that have become almost standard will begin to flutter when piston acceleration rises above about 60,000 ft/sec^2 and it is entirely too easy to exceed that limit with a modified touring engine. Therefore, I would again urge you to do your homework before starting a development program with any engine. A formula for predicting the onset of ring flutter is provided in the chapter headed, "Fundamentals", and you may save yourself a lot of grief by determining your engine's red-line with paper and pencil instead of through experimentation. At the same time, I must caution you against simply assuming that very narrow rings are an advantage in all engines. In fact, there is no detectable power difference between the standard 2.0mm ring and the "racing" 1.0mm ring below 7000 rpm, and the wider ring has the advantage of better durability right up to the point where piston acceleration starts it fluttering. Neither is there any advantage, below 7000 rpm, in the use of single-ring pistons. Above that level the lower friction of the single-ring piston begins to make a difference, but in the lower speed ranges you may as well use two-ring pistons and take advantage of their "second line of defense" capability.

Selection of ring-type will usually have been made for you by the piston manufacturer, and my advice is that you do not try to improve upon his judgment, which will be almost impossible in any case. You cannot, obviously, re-machine a piston made for 2.0mm rings to take 1.0mm rings -unless you cut a new ring groove above the existing grooves, and that would position your ring perilously close to the piston crown and almost certainly lead to immediate ring failure. The only way around this is to fit a Dykes-pattern ring, right up at the piston crown - as was noted previously. Such modifications can be very successful, if you have the right ring for the application and cut the groove correctly for the ring, but I cannot recommend the procedure simply because there is so much room for error. In general, I think it is far better to replace the stock piston with one fitted with thinner rings - even if the replacement piston is cast of somewhat inferior material, as is often the case. After all, the best of pistons will fail if its rings are not suited to the job it is being asked to perform. On the other hand, rings of less-than-desirable material will perform very well in racing applications if replaced frequently, and if they have not been crudely finished. Much of the ring's ability to function is related to this latter aspect. The ordinary cast-iron ring is fragile, and will shatter very quickly if allowed to flutter, but it will perform entirely satisfactorily if its lower surface is smooth and true, and seals against the bottom of the ring groove. Rings made of nodular cast-iron have the same wear-resistant properties, and are vastly stronger, for which reasons this material is almost universally used. Surface coatings, ranging from chromium to Teflon, are often applied to the piston's ring's face, to improve service life and /or prevent scuffing during break-in.

Ring sticking is a problem to be faced with all high-output two-stroke engines. Carburized oil may lock the ring in its groove after a remarkably short period of running if the ring lacks sufficient vertical clearance (usually, from 0.0015- to 0.0040-inch) or if the ring is located too near the piston crown. More frequently, the problem stems from the oil being used for lubrication, and it is most unfortunate that the very oils providing the best lubrication are the ones most likely to cause ring sticking. Castor-based oils, particularly, will build up thick layers of varnish inside the ring groove, unless the oil contains a considerable percentage of detergent chemicals.

Apart from the L-section Dykes ring, most piston rings have a basically rectangular cross-section, but you will find many minor variations on this arrangement.

Currently very popular is the "keystone" ring, which has a tapered section, with either the upper or lower surface, or both, sloping away from the ring's outer face. The reason for this primarily is to keep the ring and its groove scrubbed free of carbon and varnish. In four-stroke engines the rings are free to rotate, and do, and their rotation performs this scrubbing. Two-stroke engines nearly always have their rings pinned, to prevent them from rotating and the ring's ends from springing out and becoming trapped in a port. Hence, the need for some other form of scrubbing action. Seldom is the taper in a keystone-type ring more than 7-degrees, and it is all too easy to attempt installing one of them upside-down, so you should give particular attention to the ring's markings. Such markings vary in kind, but without exception they will be on the ring's upper surface.

Another point of trouble can be the ring's locating pin, and if you encounter difficulties with locating pins working loose, the source of the trouble nearly always will be in the exhaust port. The racing engine's very wide exhaust port (width representing, in extreme instances, up to 70-percent of cylinder bore diameter) leaves a lot of the ring's diameter unsupported when the piston is down in the lower half of the cylinder, which allows the ring to bulge out into the port. Making the port opening oval and chamfering its edges will prevent the ring from snagging, as these things ease the ring back into its groove as the piston sweeps back upward. However, while the ring may not snag on the

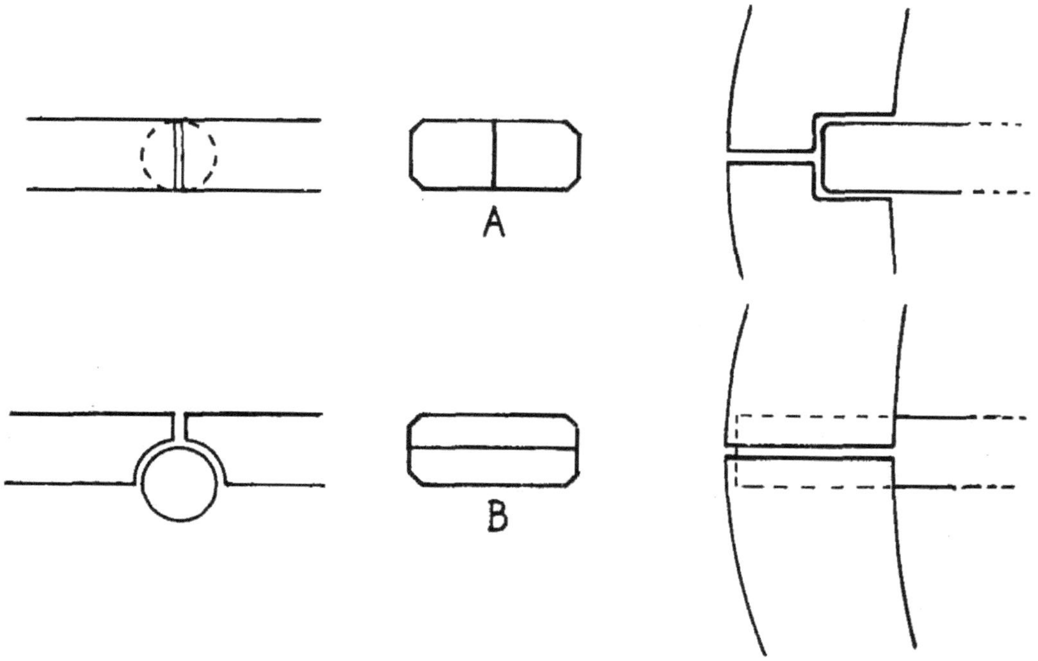

A locating pin recessed into the ring groove and covered by the ring ends (A) gives slightly better sealing than the arrangement in which half the pin is below the groove and is overlapped by the ring.

port, it does get stuffed back into its groove fairly rudely, and that may have a very bad effect on the locating pin: On most two-ring pistons, the locating pins are positioned adjacent to the areas of blind cylinder wall between the intake and transfer ports -placed about 90-degrees apart - to provide a long path for gas leakage. Thus, when the ring bulges out into the exhaust port and then is stuffed back, the end of the ring is pushed into hard contact with the pin, and after a sufficient number of hard blows (and these accumulate rapidly at, say, 10,000 rpm) the pin begins to loosen and it will gradually enlarge the hole in which it is inserted enough to work completely loose. Then the ring is free to rotate, and it quickly works its way around to catch the end in a port. At risk of seeming immodest, I will admit to having isolated this problem for Yamaha several years ago and today that firm's racing engines have pistons with locating pins positioned 180-degrees from the exhaust port. Touring engines, which have much narrower exhaust port windows and thus treat their rings more gently, usually benefit from having their two rings' end-gaps placed more nearly on opposite sides of the piston, as described before.

In some racing applications, the standard rings are adequate to the engine speeds anticipated, but overall performance may dictate a much wider-than stock exhaust port. Then, the "offset" ring-locating pin may prove prone to precisely the sort of loosening and subsequent failure described in the preceding paragraph, which will lead you into a piston modification that can be very tricky: installing a new locating pin in the back of the ring groove. This gets tricky because in many cases the pin will be half-in, half above, the ring groove and it is impossible to drill the hole for a new location after the groove is machined. Impossible, unless you cut a small piece of aluminum to exactly fit the ring groove, filling it flush, in which case you drill your hole half in the piston and half in the filler piece. Then you remove the filler and your hole is ready for the pin - which introduces yet another problem: what to use for a pin? Steel wire is a good choice on grounds of strength, but is likely to work loose simply because the aluminum in which it is pressed grows and contracts so much with changes in temperature. A small-diameter "split pin" (which is like a rolled tube) is a better choice, but if suitable sizes are not available, then n pin made of hard brass is at least as good.

WRISTPIN/CRANKPIN BEARINGS

Back in the days when pistons were uniformly poor and two-stroke engines wouldn't be run very fast, wrist pin bearings were almost always a simple brass bushing. Such bushings work very well in four-stroke engines, but lubrication is much less lavish in the crankcase-scavenged two-stroke and added difficulties are created by the essentially uni-directional loads placed upon it, which prevent the piston pin from lifting away from the lower part of the bearing and admitting oil to the load-carrying surfaces. For those reasons, the plain bushing has now almost universally been replaced by "needle" roller bearings, which are more easily penetrated by such oil as is available and in any case need much less oil. This last is of very particular importance in high output engines, as the heat flowing down from the piston is certain to thin any oil present to a viscosity approaching that of water. But all these difficulties not withstanding, the needle-roller bearing is wonderfully trouble-free, and if you encounter problems at the hinge between the connecting rod and piston pin, those problems will almost invariably be with breakage of the bearing cage. Given the extremely low rotational speed of the bearing in question, no cage is really needed except to make engine-assembly easier: the cage holds all the needle-rollers in place while the piston is being fitted to the connecting rod. The arrangement certainly makes working on the engine less complicated, but as it happens, the cage becomes the bearing's weakest link. Piston acceleration at high speeds is also applied to the bearing cage, and it may shatter under the strain - which sends a shower of particles from the broken cage and loose needles down into the crankcase. The debris thus liberated invariably gets pumped up through the transfer ports, into the cylinder, and more often than not a roller will get trapped hanging half out of a port by the piston with dire consequences to both.

Yamaha's TD1 was particularly prone to small end bearing cage failures, and I learned the hard way to replace these bearings if I saw over 11,000 rpm on the tachometer even for a moment, for their cages required only a moment's battering before cracks would start to spread and outright disintegration soon followed even if I indulged in no more excursions past the red-line. This difficulty has been overcome with cages made of tougher material; it is possible to accomplish the same thing by using crowded needles and no cage at all, which does require that a washer be fitted on each side of the connecting rod, to take up clearance so that the rollers cannot escape. Getting the thing assembled (with the roller glued in place with grease) is enough to make strong men weep with frustration, but it absolutely insures reliability at this point in the engine and is *a* measure worth remembering if problems with broken wrist-pin bearing cages do occur.

McCulloch, the chain-saw people, have used an arrangement similar to the one just described for years, but they have reasons other than simply working around bearing cage failures at the wrist-pin end of the rod. It was discovered at McCulloch that failures at the crankpin bearing were traceable to the thrust washers most manufacturers of two-stroke engines use to center the rod on the crankpin. These washers usually are made of brass, or steel with a copper coating, and they do not find high rubbing speeds and scanty lubrication at all agreeable. At very high crankshaft speeds, they register their protest by overheating, and this causes a rise in temperature all around the connecting rod's big end,

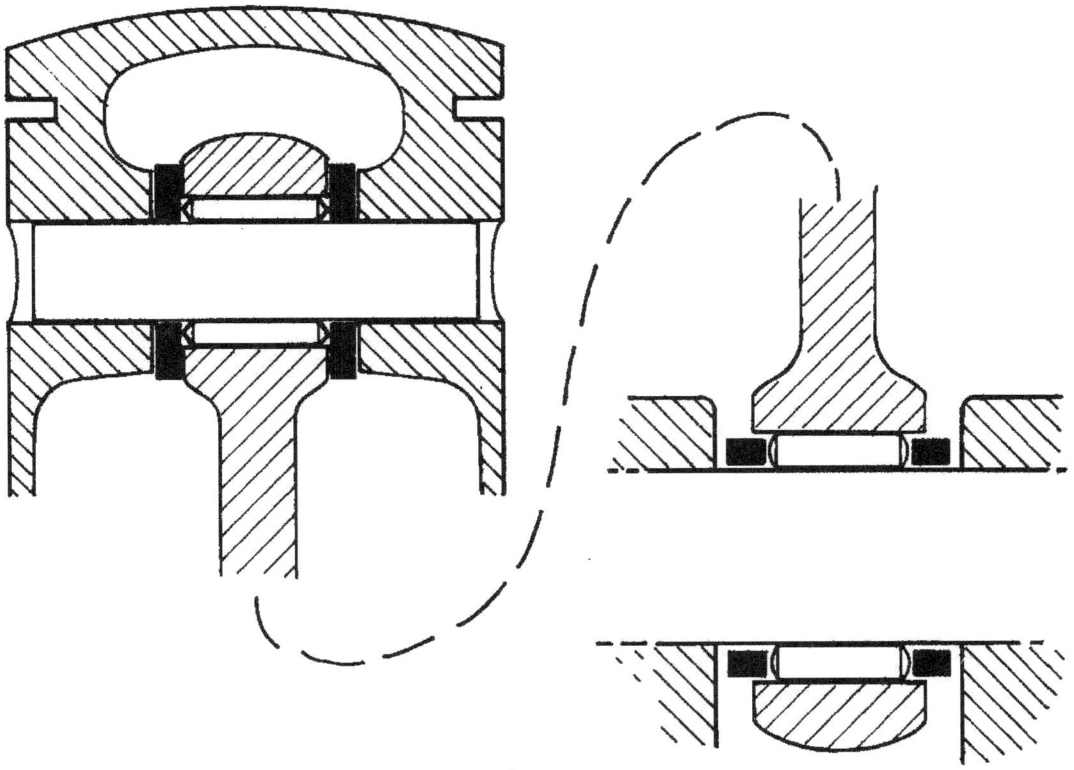

Difficulties in assembly have prevented wide use of encaged rollers and retaining / thrust washers at the wrist pin, but the arrangement reduces friction and improves oiling down at the critical crankpin.

which thins the oil present enough to create yet more friction, more overheating, until at last the thrust washers, roller bearing and cage are hot enough to “flash” the oil. At that point, lubrication is nil and friction quickly melts the bearing cage and wears flats on the rollers. McCulloch's engineers reasoned that the point of failure could be pushed upward materially simply by removing the thrust washers, which is what they did. Of course, the connecting rod still had to be centered over the crank, but this task was given to a pair of thrust washers up inside the piston. The improvement in terms of elevating the McCulloch kart engine's maximum crank speed was in the order of 1500 rpm, and it is worth noting that Yamaha borrowed this idea for use in the 17,000 rpm GP engines the company raced in 1968. It is interesting that in those engines, the piston rings were only 0.6mm in thickness.

Crankpin bearing failures also stem from the use of excessively heavy bearing cages. Sheer rotational speed is not enough to burst a cage of such small diameter and mass, but the fact that the cage must accelerate and decelerate, relative to the crankpin as the connecting rod swings, will cause difficulties unless the bearing cage is very light. In effect, the rollers must push the cage up to speed and then slow it, and if the cage has enough inertia it will resist this pushing and pulling enough to skid the rollers - at which

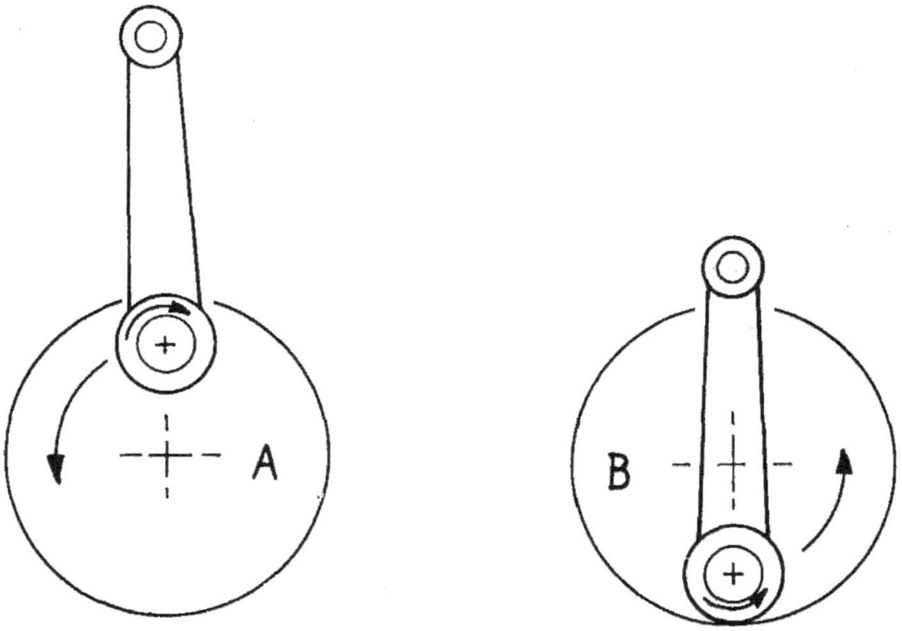

Connecting rod swing effectively increases crankpin bearing rotational speed at TDC (A) and slows it at BDC (B); this speeding and slowing tends to make the bearing rollers skid at high crankshaft speeds.

point they momentarily become a plain bearing- a job for which they are poorly constituted. The skidding rollers generate a lot of heat, through friction, and the heat leads the bearing into the same deteriorating cycle to outright failure as was outlined for the thrust washers. Most modern engines have steel crankpin bearing cages, copper- or tin-plated to provide a low-friction surface to bear against the rollers, crankpin and connecting rod eye. These replace the phosphor-bronze cages of the recent past - which replaced the inelegant aluminum and brass cages of a yet-earlier era. But the best current "big-end" bearing cages are made of titanium and silver-plated. Experimenters with near-unlimited funds may like to try titanium bearing cages, but when having them made they should know that the bearing retaining slots must be machined with edges parallel to within 1/200 with each other and with the crankpin (assuming a parallel condition between cage and crankpin axis). It is not a job for someone with a bench-vise and a file. On the other hand, if employing silver-plated titanium cages and moving the thrust washers from the crankpin to the piston will elevate your engine's red-line by 2000 rpm, then they clearly will pay dividends in horsepower - if port-timing, etc., is adjusted correspondingly.

Connecting rods should not be lightened, or even polished, unless you intend going all the way in this direction and will finish the job by having the part shot-peened. Forgings acquire a tough skin in the process of being pounded into shape, and I know of instances where connecting rods that were entirely satisfactory in standard condition promptly broke after having been polished. I do think, on the other hand, that there is a margin of safety to be gained by smoothing off the rough edges where the flash has been sheared away from the forgings. Notches are, in the engineer's language, “stress raisers” and you can do the connecting rod no harm in removing them. Lightening the connecting rod is, however, a poor choice of ways to use one's time, because a rod intended for the loads at, say, 8000 rpm is going to be overstressed at 10,000 rpm and if anything, material should be added to the rod, not removed. On the other hand, one sometimes can improve bearing reliability by opening slightly the oil channels at the ends of the connecting rod. I do not recommend that you actually cut into the bearing surface, but oil delivery to the bearing will be improved by tapering the entry. *Do not* extend the taper all the way to the bearing surface, as the sharp edges thus formed will flake away as the engine runs and cause a bearing failure.

Crankshaft main bearings seldom are troublesome, except in engines that have

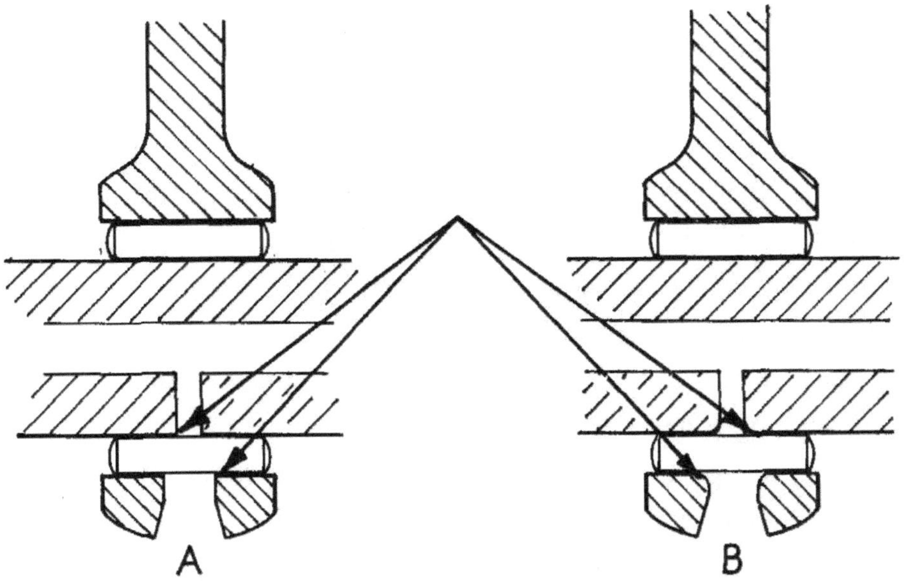

The sharp corners at the rod's oiling slot and around the crankpin oil feed hole (A) are subject to flaking that can cause a bearing failure. Small radii ground at these points prevent such problems.

been in storage for a long time and have had corrosion at work in these bearings -or unless the bearings have been mishandled. Bearing steels are very tough, but you definitely can pound small pits in the races by injudicious use of a hammer, and pits also can be formed by rusting. Bearings damaged in either fashion should be replaced, as the pits will soon spread and become minor trenches, as a result of an activity called "Brinelling", which actually is a form of work-hardening. The bearing's rollers and races have casehardened surfaces, but the metal under this thin case is relatively soft, and it is compressed and released (at any given point) as the bearing turns under a load. If the load is high enough, or the bearing in service long enough, the repeated compressions will literally fatigue the metal, and tiny particles of the surface will start flaking away - which becomes visible as the "tracking" seen in the races of a worn-out bearing. Any bearing will start flaking at some point in its life; bearings with races damages by rust, etc. will begin such flaking almost immediately. Incidentally, in very highly loaded bearings the flaking may be started by the sharp edges around *any* interruption in the bearing's surface, if the rollers pass over those edges. Oiling slots in the rod's big-end are prone to develop this kind of failure, and the same sort of flaking is sometimes observed around the oil feed holes in the crankpins of engines equipped with "direct-injection" oiling systems, like the Suzuki's and Kawasaki's. Remove the sharp edges, and you remove the problem - if any. There is sufficient margin of strength in stock production engines so that the problem does not occur; you may find it in the course of reaching for crank speeds substantially above the stock specification.

Somebody is always telling me about having an engine "balanced", and I always smile nastily when the engine in question has fewer than four cylinders. In point of fact, the single-cylinder motorcycle engine cannot be brought into dynamic balance, for if you counterweight the crankshaft to compensate for the full weight of the piston and rod, you will simply have moved the shaking force from being in-plane with the cylinder axis 90-degrees. "Balancing" one of these engines consists of finding a balance factor, in percentage of reciprocating mass, which is kind to the engine's main bearings and does not excite resonance in the motorcycle's frame. In-line twin- and three-cylinder engines always have a rocking couple. By and large, the stock crankshaft counter-weighting will be correct for most applications, and unless you want to get into a really lengthy experimental program there is nothing to be gained in making changes.

CRANK ASSEMBLY

There are gains in power and reliability to be had from carefully aligning your crankshaft and main bearing bores, and in getting the cylinder axis precisely perpendicular to the crankshaft. As it happens, there is more variation in production tolerances when the various parts of a crankshaft are made than can comfortably be tolerated in a racing engine. Crankpin holes in flywheels are not all precisely the same distance from the main shaft axis; factories "select-fit" these parts, and you can be fairly certain that a new crankshaft is true, but if you manage to ruin any of its flywheels, do not assume that a replacement flywheel, selected at random from the nearest parts bin, will be a *satisfactory* replacement. Crankpin holes, in facing flywheels, should be matched to within 0.0002-inch with regard to their offset from the main shaft. If your local source cannot supply a single replacement wheel within that tolerance limit, I strongly urge that you purchase a complete, new crankshaft - with flywheels matched at the factory. And when rebuilding a crankshaft, with new crankpins and bearings, be certain that it is aligned to at least the tolerances suggested by the manufacturer's workshop manual. Also, check your crankcases for main bearing-bore alignment - and, more important yet, that the cylinder is exactly perpendicular with the crank axis, for any tilting will be reflected in added friction in the hearings (especially at the thrust washers) and in the piston itself.

Do not attempt to second-guess the manufacturer with regard to crankshaft and crankpin bearings unless you have very specialized knowledge in this field or can obtain the advice of someone who is an expert. Main bearings, particularly, should not be replaced with just anything that will fit, as n very special kind of bearing is employed in these applications, with clearances to accommodate the expansion and contraction of aluminum bearing housings. And the same cautionary note must be added with regard to crankshaft seals, which in the high-speed, two-stroke engine must survive extremes in temperatures and rotational speeds with very scanty lubrication. Not so very long ago, seal failures were common, but now that means have been found to Teflon-coat seal's lips, trouble is usually encountered only when the seals have been damaged in the course of installation. So handle the seals carefully, and pre-coat them with a good high-temperature grease before assembling your engine. You can also improve their reliability somewhat by polishing the area on the main shafts against which they hear to a glassy finish. The seals themselves will polish the shaft eventually, but at considerable expense to their working life.

By and large, problems with piston, connecting rod bearings, crankshaft and seals can be avoided simply by following the recommendations made in the manufacturer's shop manual. The single exception to this is in the fit between piston and wristpin, for the very high temperatures in a modified engine tend to cause a breakdown in the lubrication between pin and piston. Trouble can be avoided in the racing engine if the wristpin is a light, sliding fit through the piston; it should slip through of its own weight, without forcing, for if it is tight enough so that you have to tap it through with a mallet, you eventually may have to remove it with a hydraulic press. Too-light fits may be corrected by using an old wristpin as a lap, and a dash of some fine, non-imbedding lapping compound to polish out the piston's pin-bore to size.

CYLINDERHEADS

For the Otto-cycle engine, of which the two-stroke is an example, there is a theoretical level of efficiency, in terms of converting heat into work, referred to in basic engineering texts as "air standard efficiency". In this, it is assumed that the cylinder is filled only with dry air, and heat then added, which ignores the fact that in practice the air contains some moisture and a considerable percentage of hydrocarbon fuel. Even so, this theoretical level of efficiency, calculated against compression ratio, provides a useful yardstick against which actual efficiency can be measured - and it tells us a lot about the effects, on power output, of compression ratio. For example, at a compression ratio of 5:1, air standard efficiency is 47.5-percent, while at 10:1, it is 60.2-percent. That is, of course, a very great gain, and the consequences - measured at an engine's output shaft-are the reason for many experimenters' fixation on "raising the compression". Certainly, increases in compression ratio, which may be accomplished simply by trimming a few thousandths of an inch from the cylinder head's lower surface, can work minor miracles with an engine's performance.

But higher compression ratios can also bring about a mechanical disaster: improvements in power gained in this manner are purchased at a disproportionate cost in peak cylinder pressure, leading to reduced bearing life and sometimes to an outright failure of a connecting rod or crankpin. Moreover, because the higher pressures are reflected in a proportionately greater side thrust at the piston, frictional losses are such that net power gains are always less than the improvement one would expect from the calculated air standard efficiency. Finally, heat flow from the combustion gases into the surrounding vessel (piston crown, cylinder head, and cylinder walls) rises increasingly sharply with compression ratio, so that a number of thermal-related problems intrude into the already complicated relationship between compression ratio and power.

The worst of these problems is the overheating of the piston crown. A too high compression ratio will raise piston crown temperatures to the point where heating of the mixture below the piston, in the crankcase, reduces the weight of the charge ultimately trapped in the cylinder during the compression stroke to such extent that net power suffers -no matter what Mr. Otto's air standard efficiency formula may say. And if the compression ratio is high enough, heat input into the piston may raise the crown temperature to the point where detonation and then pre-ignition occur. These phenomena will, in turn, very quickly further raise piston crown temperature to such extent that the piston material loses enough of its strength to yield to the gas pressure above – the piston crown then becoming either concave (which drops the compression ratio to a tolerable level) or develops a large hole (and that reduces the compression ratio to zero:zero).

Many people have encountered this last effect, and the tuner's one-time favorite ploy of "milling the head" has fallen into disrepute. But it also is possible to encounter trouble without recognizing it: There is a delicate balance between gains from increased compression ratios and losses due to increased temperatures -which appear not only at the piston's interior, but also throughout the crankcase, crankshaft, rod and all the rest of the engine's interior contacted by the air/fuel mixture. When these parts are hotter, the mixture's temperature is also raised, along with its free volume. Thus, the mixture's temperature-induced efforts to expand inevitably force part of it out the exhaust port, and as power is related very closely to the weight of the charge captured in the cylinder, this heating shows up as a power loss. The trick is to balance crankcase heating and compression ratio. There is an optimum combination for every set of conditions, but finding that optimum without heat-sensing equipment and a dynamometer is exceedingly difficult.

THE COMBUSTION PROCESS

Not too surprisingly, the equilibrium described is influenced by combustion chamber design-as is the point at which smooth burning gives way to the outright explosions we call detonation. This aspect, too, is widely appreciated, but not widely understood. In truth, most people have very little understanding of the events that follow ignition; events that are highly complex if studied with regard to their chemistry but really quite straightforward taken in less narrow terms. Much of the misunderstanding that exists has been created by the popular press, which insists upon saying that a piston is driven downward on its power stroke by a burning mixture. In reality, the burning of fuel in the cylinder is simply a means of raising the temperature of the working gas (air; actually a mixture of gases) and thereby raising its pressure. This relationship was formulated long ago by Boyle as:

$$\frac{P_2}{P_1} = \frac{T_2}{T_1}$$

Where, of course, P is pressure and T is temperature. The whole business gets complicated in the internal combustion engine by the changes in the cylinder's contents due to the combination of elements in the working gas with fuel, but it still basically is a case of raising the working gases' temperature and thus raising their pressure, and it is *that* which pushes the piston down and makes the horsepower. In fact, burning will have been all but completed by the time the piston starts downward on its power stroke.

Here, for anyone who cares, is what happens from the moment of ignition: Several thousandths of an inch of travel before the piston reaches the top of its compression stroke, representing somewhere between 20- and 45-degrees of crank rotation, the trapped air/fuel charge is ignited by the spark plug and burning commences. At first, the process proceeds quite slowly (relative to subsequent crank rotation before

TDC). A small bubble of fire expands gently away from the point of ignition between the spark plug's electrode and ground wire, and if all combustion were to continue at this pace it would hardly be completed in time for the following compression stroke. However, this small flame quickly heats the remaining mixture enough to enormously increase the rate at which burning occurs, and after the initial delay, the flame-front accelerates outward from its point of origin with ever-increasing rapidity - sweeping throughout the combustion chamber. And if the engine has been given the proper amount of spark advance, the piston will have just moved up to the top of its stroke as the rapid phase of combustion begins, so that the bulk of the burning is done while the piston is virtually stopped and the mixture compressed to minimum volume. By the time the crankshaft has rotated a few more degrees, and the piston is once again moving downward, the combustion process will have been almost entirely completed.

The preceding is what happens in the normal course of events; combustion does not always occur that neatly. The most common, regrettable combustion irregularity is detonation, the harsh knocking you hear just before an engine seizes, or melts a piston - and the noise you would hear, when running an engine on a dynamometer, as the needle on the scale begins an ominous retreat. Unhappily, the very process by which the mixture in the combustion chamber is re-heated before its actual contact with the flame-front advancing from the spark plug, and rapid combustion thus made possible, is the process that may also lead to the sudden explosion of the combustion chamber's contents that we call detonation. Here's how it happens: It has already been noted that as the flame-front advances, the combustion chamber's remaining unburned mixture is heated, and this heating is caused not only by direct contact with the flame, but also by radiation and the overall pressure rise within the chamber. If the temperature of this remaining mixture is raised to its ignition point, all of it is consumed at the same instant in a single explosion. This explosion creates a shock, due to a fantastically rapid pressure rise, that strikes out against all its surroundings hard enough to make detonation's characteristic knock - and it is a shock with a force often sufficient to break the spark plug insulator's tip and damage both the piston and bearings. Even so, its worst effect is to force a lot of heat out into the piston, cylinder head and the cylinder walls. These are thus brought to abnormally high temperature, which tends to overheat the next air/fuel charge and make it detonate even more quickly and severely.

Should this detonation continue, it will overheat the engine's upper end to the point where ignition occurs before there is a spark: compression heats the mixture in any case, and when a lot more heat is added from the piston crown, etc., the mixture will be

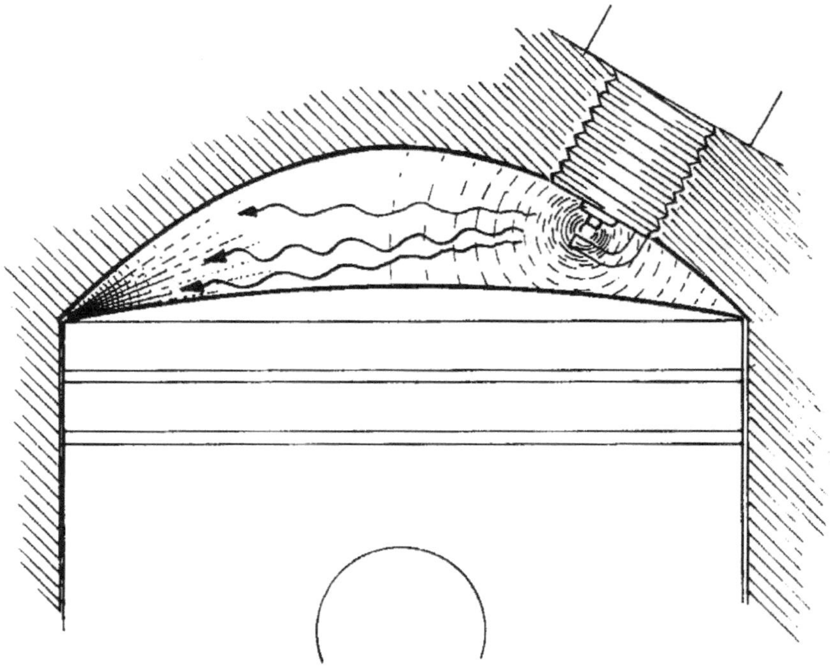

In this combustion chamber there is no squish band and the spark plug is off-set. Detonation will occur when pressure, and heat radiating ahead of the flame front, cause the end gases to ignite.

brought to "pre-ignite". Detonation has a very bad effect on power output; pre-ignition (thought by some to be the same phenomena) is even worse in that regard, but will not long continue unnoticed as it will very rapidly overload the piston - in both the thermal and mechanical sense - beyond the point of failure. Knowing that, you will appreciate that detonation is to be avoided if at all possible. One way to avoid detonation would be to simply hold the compression ratio to some very low number, as they would reduce the pre-combustion temperatures and thereby make detonation unlikely if not impossible. But that method is mostly (the exception I will deal with shortly) too expensive in terms of power-output efficiency. A better method is one employed in most engines today: use of a "squish type combustion chamber, in which the mixture is trapped in a small pocket under the spark plug, and the rest of the cylinder head surface over the bore is made to fit closely against the piston crown when the piston is at top center.

We have England's Harry Ricardo to thank for this type combustion chamber, which he created to cope with conditions that ceased to exist long before most of us were born. During the conflict that wracked Europe just after the turn of this century, there were not only shortages of internal combustion engine fuels, but the fuels available were of very poor quality – and would detonate severely in the side-valve engines of that

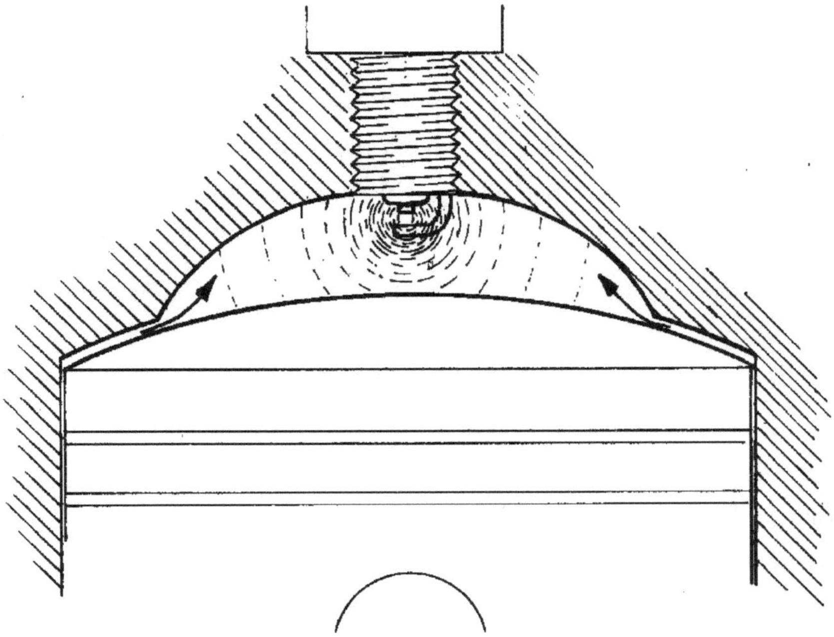

Combustion chambers with centrally located spark plugs and squish bands give a short flame travel in a compact mixture pocket; the end gases, being in thin layers, are too rapidly cooled to ignite.

period unless the engines were operated with a much-retarded spark, or their compression ratios lowered to about 4:1, or both. These measures had a terrible effect on fuel economy, naturally, and the problem led Ricardo to do serious research into the nature of detonation. We now know that the side-valve engine is particularly prone to detonation, as it of necessity has a very long combustion chamber. Ignite a fire at one end, and it will be a long while reaching the far corners of the chamber. In the interval between ignition and the completion of burning there is ample opportunity for the unburned part of the charge to overheat and ignite.

SQUISH BANDS

Ricardo solved the problem, once he had determined its nature, by lowering the underside of the cylinder head in that part of the chamber over the piston. Thus, most of the mixture was concentrated right at the ignition source, and would be more likely to burn without detonating. The small part of the mixture caught between the cylinder head's squish band and the piston was still subject to compression heating, but was fairly effectively shielded from radiation and was, moreover, spread in such a thin layer that it would resist ignition from any cause - as it would lose heat into the relatively cool piston and cylinder head too fast to ignite.

That still is the secret of the squish-type cylinder head: It concentrates the main charge in a tight pocket under the spark plug, and spreads the mixture at the cylinder-bore's edges too thinly to be heated to the point of ignition. These "end gases" do not burn with the main charge, and are only partly consumed as the piston moves away from top center and releases them from their cooling contact with the surrounding metal. And right there is the disadvantage that comes with the squish-band cylinder head, for mixture that does not burn is mixture that contributes nothing to power output. Of lesser importance, though only in this context, is that those end-gases contribute heavily to the release of unburned hydrocarbons out the exhaust pipe and into the atmosphere, and for that reason automobile manufacturers are now relying much less heavily on squish-band chambers for combustion control. You may be interested to know, too, that in many cases a non-squish combustion chamber, with its complete utilization of the mixture to offset the power-limiting effects of a necessarily-lower compression ratio, has proven to be best in absolute terms of power and economy. McCulloch, for example, make engines with both squish and non-squish cylinder head configurations - having found that both have their applications.

Our application here, of course, is strongly biased toward maximum horsepower, and that points toward a squish-band head - which is what you will have in most motorcycles in any case. I will warn you, now, that it may be unwise to follow the old-time tuner's habit of increasing an engine's compression ratio as an opening gambit in the quest for better performance. Indeed, before your work is done you may find it necessary to reduce your engine's compression ratio *below* the stock specification. You see, in the final analysis it is not so much compression ratio as combustion chamber *pressure* that determines the limit - and these are not at all the same things. Your stock engine, with a carburetor size and porting chosen to lend it a smooth idle and easy starting, does a much less effective job of cylinder-filling than will be the case after it has been modified. More important, it will probably have an exhaust system that has more to recommend it as a silencer than as a booster of horsepower. These factors, in combination, make a very great difference between the cylinder pressures at the time of ignition in the stock and modified engine. Even given a certain willingness on your part to use a fairly cold spark plug - changing it frequently - and a further willingness to replace pistons and bearings more often in payment for added power, it may *still* be necessary to stay with the stock specification for compression ratio. Or, as I have said, to lower the engine's compression ratio from the stock condition. This last will be particularly true if you succeed in creating a much better than stock exhaust system.

CYLINDERHEADS

By and large, you would be well-advised to ignore the whole business of compression ratios in favor of cranking pressures. There is, after all, a big difference between the kinds of numbers you get by performing the traditional calculations to find compression ratio, and what is happening as the engine turns. My experience has been that you can use cranking pressures of 120 psi without worrying much about overheating anything. Maximum power will be obtained at cranking pressures somewhere between 135 and 165 psi. Going higher with compression, in a conventional motorcycle engine, can give a neat boost in low speed torque, but the thermal repercussions of higher cranking pressures will surely limit maximum output. On the other hand, fan-cooled kart engines perform very well at cranking pressures up at 200 psi, and water cooled engines behave much the same.

One of the most undesirable side-effects that comes with too-high compression ratios is an enormous difficulty in getting an engine to "carburet" cleanly. When the compression ratio is too high, you'll find that an engine's mixture-strength requirement has a sharp hump right at its torque peak that no motorcycle carburetor can accommodate. You'll realize, after working with high-output two-stroke engines, that all of them are to some degree liquid-cooled - and that the cooling liquid is gasoline. It is

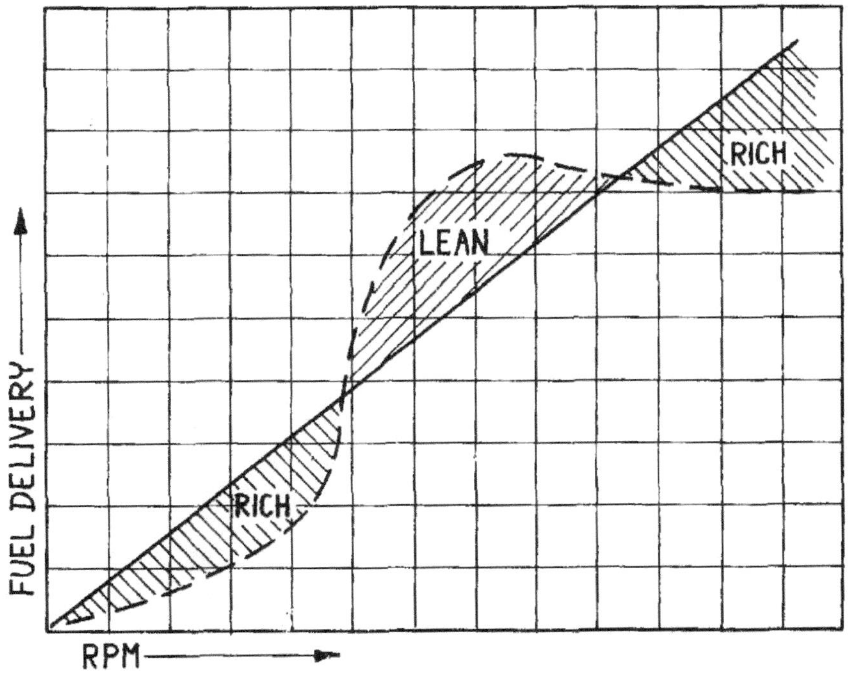

Carburetors produce a mixture of nearly constant strength, but an expansion chamber's boost capability varies with engine speed and may alter the engine's mixture requirement above and below what the carburetor can actually deliver.

true that an over-rich mixture tends to dampen the combustion process, and reduce power, but here again we find ourselves faced with the necessity for finding a balance between evils: We have overheating to rob power on one side, and we can cool the engine with gasoline, but too much fuel also robs power. The solution is a beggar's choice, in which we try to find the cross-over point between overheating and over-rich mixtures.

In an engine intended purely for road racing, with a torque peak virtually coincidental with its power peak and driving through a very close-ratio transmission (enabling the rider to hold engine-speed within narrow limits), making this beggar's choice is a fairly straight-forward proposition: you play with jetting until the motorcycle runs fast. However, road racing conditions allow you to stay right on the mixture-requirement hump; you don't have to worry about what happens two-thousand revs below the power peak, because that's below what you'll use in a race. Motocross racing is another matter entirely, and an engine with a mixture-curve hump will drive you absolutely mad. Jet a motocross engine so that it doesn't melt a piston every time it pulls hard at its torque peak, and (if its mixture-curve is humped) it will be huffing soot and losing power above and below that speed.

The answer to this problem is to iron out that mixture-requirement hump, because no matter how much work you do with the carburetor, it never will be able to cope with the engine's needs. All the carburetor knows, really, is how much air is moving through its throat, and it adds fuel to the air in proportion to the rate of air-flow; don't expect it to know when the piston is getting hot and respond by heaving in some more fuel. How do you get rid of the hump? You do it mostly by substituting a somewhat less effective expansion chamber: one that gives more nearly the same boost all the way through the speed range you are obliged to use by racing conditions, without any big surges. That will result in a drop in peak power, obviously, but you can compensate for it to a considerable extent with the higher compression ratio you previously were forced to forego in the interest of keeping the piston crown intact when the expansion chamber did its big-boost routine. Again, it is all a matter of finding the balance.

No matter what the compression ratio you ultimately use, it will have been influenced much more than you probably suspect by the combustion chamber configuration, and by certain gross characteristics of the head itself. Over the years, I have seen the fashion in combustion chamber forms swing back and forth, hither and yon, with first hat-section chambers in favor and then trench-type chambers, and torus-type chambers and so on and so forth ad *infinitum.* I was not, and am not, impressed. Combustion chamber form should be established with an eye toward only a very few special considerations, and these cannot account for even half the chamber shapes I have seen. Listed, though not really in order of importance, these are: surface / volume ratio; spark plug location; thermal loadings; and combustion control. We will consider each of these in turn.

Surface to volume ratio is important because even in the part of the combustion chamber fully exposed to the advancing flame front, there will be a mixture layer adhering to the metal surfaces that does not burn. These layers, like that trapped within the squish band, are cooled by their proximity with the cylinder head, or piston, and simply never will reach ignition temperature. And, like the end-gases from the squish band, they eventually find their way out the exhaust port, having taken no part in the conversion of fuel and air into horsepower. Thus, the best combustion chamber shape - taken strictly from the standpoint of surface/volume ratio - would be a simple spherical segment sweeping in a continuous arc from one side of the cylinder bore to the opposite side. No tricky changes in section, no squish bands, no nothing. And that is, in point of fact, precisely the shape employed in nearly all non-squish cylinder heads.

But if you want to use a true (measured from exhaust-closing) compression ratio much over 6.5:1, on a high-output engine, combustion control beyond that afforded by a non-squish cylinder head will be necessary. Considerable variation is possible, but a

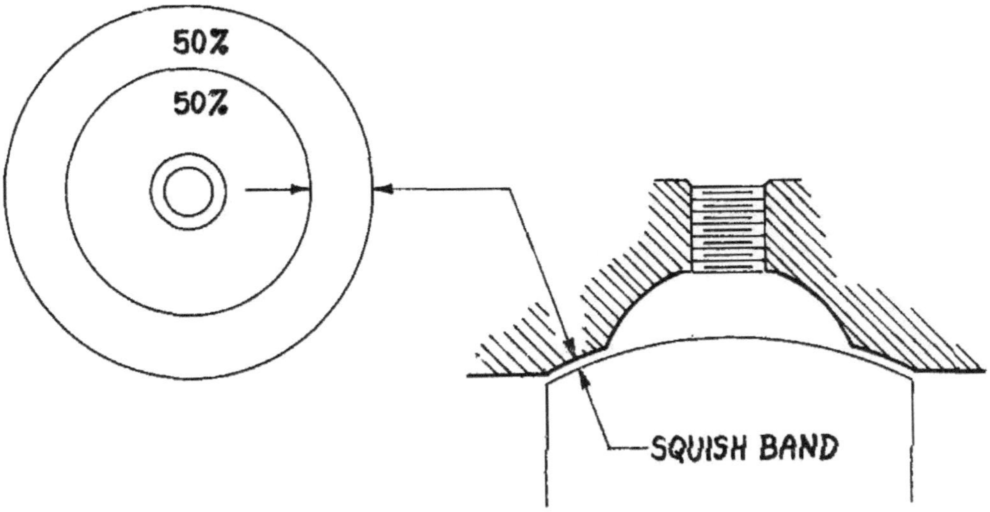

Squish bands should constitute about half the cylinder bore area. Clearance between piston and cylinder head should be held to a minimum to avoid effectively losing about 5% of working mixture.

good rule to follow is to make the cylinder head's squish band about 50-percent of the cylinder bore area. For example, in a 3-inch bore -which has a total area of 7.07-$inches^2$ the squish band would be wide enough to represent an area of just about 3.5 in^2. Assuming that you have centered the combustion chamber proper on the bore axis, then your squish band would be a ring having the same outer diameter as the bore, and an inner diameter of just over 2-inches. The combustion chamber itself, to meet the previously-stated minimum surface/volume requirement, would again be a spherical segment - with a radius that provides the total volume, added with that from the clearance space between piston and squish band, to give the desired compression ratio.

The clearance space between piston and cylinder head must be enough to avoid contact at high engine speeds, yet close enough to keep the mixture held there cooled during the combustion process. This vertical clearance between squish band and piston should not be greater than 0.060-inch, and it is my opinion that the minimum should be only barely enough to prevent contact -usually about 0.015-inch in small engines (with tight bearings and cylinder/rod combinations that do not grow, with heat, disproportionately) and up to about 0.045-inch in big engines.

Some disagreement exists as to the validity of claims that the squish band aids combustion by causing turbulence in the combustion chamber as a result of the piston "squishing" part of the charge between itself and the head. I don't know about that, but I do know that holding squish band clearance to a minimum means that there will be the smallest volume of end-gases escaping the combustion process, and that can be more important than you might think. For example, a 250cc cylinder with a full-stroke compression ratio of 10:1 will pack its entire air/fuel charge into a volume of only 28cc by the time its piston reaches top center. Assuming that it has a 3-inch bore, and a 50-percent squish band with a piston/head clearance of .045-inch, then the volume of the charge hiding in the squish area will be in the order of 2.6cc, or almost 10-percent of the total. That can be reduced to 5-percent merely by closing the squish band's clearance to 0.020-inch - and you'll never find an easier 5-percent horsepower difference. True, the difference measured at the crankshaft might prove to be more like 2-1/2-percent, but the addition of those small percentages can make a very large final difference.

PLUG LOCATION

Tests have shown that the best location for the spark plug is, by and large, squarely in the center of the combustion chamber, and with its gap as close to the center of the volume of trapped mixture as possible - which is logical, as that position provides the shortest flame travel in all directions. However, a number of other considerations do intrude. First, the plug gap will necessarily be at the periphery of any partly-spherical chamber, and not at its center, and trying to form a knob in the chamber roof - to move the plug deeper into the mixture volume -will upset the chamber's surface/volume ratio. Secondly, moving the plug too close to the piston seems to cause a local overheating of the piston crown, which can impose an unnecessarily low ceiling on compression ratio.

This last consideration has, in many instances, led development engineers to use combustion chambers with forms that allow the plug to be positioned well away from the piston: modified spheroids; conical sections, etc. Also, chambers with higher roofs (like those shaped as cones) with their spark plugs up at the top and the broader base down at the piston, provide a slightly slower pressure rise as combustion progresses, and are in consequence a bit more kind to bearings. Other switches in plug location may be made in the interest of easing the job of plug replacement: it is difficult to change a plug centered in the cylinder head when the bottom of a fuel tank, or frame tube, is directly overhead.

Fortunately, most engines usually are relatively insensitive to plug location as long as the gap isn't moved too close to the piston. Which raises an interesting point: The common practice of shaving material from the cylinder head's lower surface not only raises the compression ratio, and thus the thermal load on the piston, but it brings the plug gap close to the piston crown -compounding the problem. A better approach to obtaining

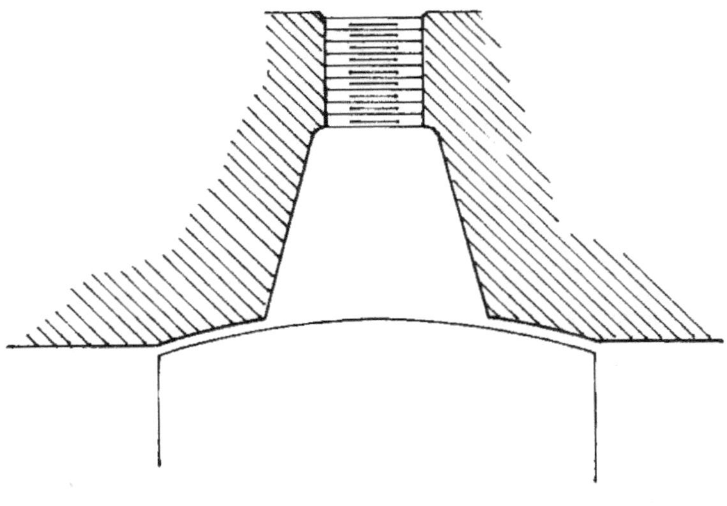

Conical combustion chambers reduce thermal loads on the piston and mechanical loads on the rod and bearings, but have poor surface / volume characteristics and give a too-broad squish to be efficient.

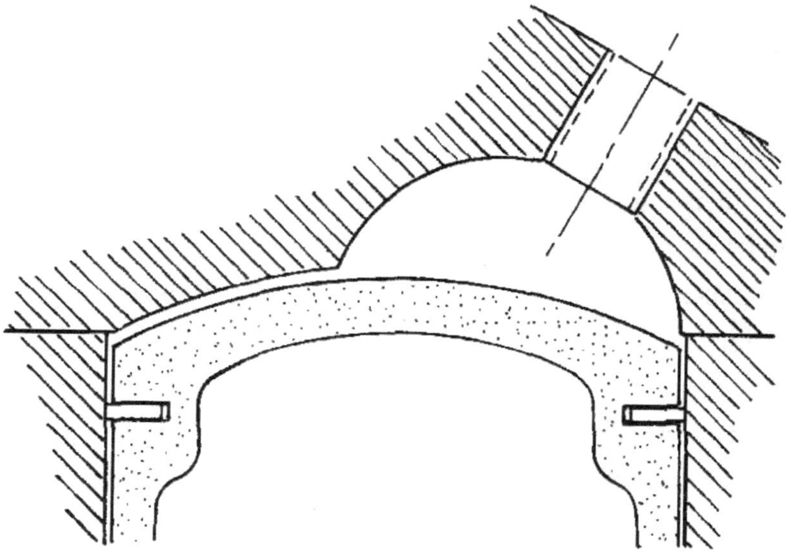

Heat input into the piston tends to be higher on its exhaust-port side. Moving the combustion chamber over to the intake side will help balance the thermal load and permit higher compression ratios.

increases in compression ratio is to purchase a cylinder head developed to do the job properly. Yamaha's GYT-kit heads, for example, provide the right compression boost, have their spark plugs properly located, etc. Other made-for-the-job cylinder heads offer the same fundamental advantage, which is that you get to buy a lot of other people's engineering at a very low cost.

Not all cylinder heads have their spark plugs and combustion chamber pockets centered over the cylinder bore, and there are good reasons for most of the variations in form one sees in the products of the major manufacturers: For instance, piston crown temperatures seldom are even, and while the overall temperature distribution pattern is understandably inclined toward maximums in the center of the crown, circumstance can also lend a bias toward the exhaust port. That bias comes not from any heat-input pattern, but rather from the manner in which the piston crown is cooled- by heat transference into the air/fuel mixture below, and into the piston skirt, from whence it is transferred out into the cylinder walls. Cooling provided by the turbulent crankcase charge is more or less even; the same cannot be said of heat losses into the cylinder, for the temperature gradients around the cylinder's walls are most uneven. The area around the exhaust port is hotter than that back at the intake port, even though the exhaust-side of the cylinder is in most instances the recipient of the direct cooling-air blast. Moreover, the exhaust-port side of the piston skirt is bathed in fire every time the port opens at the end of a power stroke. The overall result is to move the maximum temperature point on the piston crown toward the exhaust port.

Now, when that maximum temperature bias begins to seriously overheat the side of the piston, you are likely to see some severe piston ring problems develop: Too-high temperatures will eventually be a disaster for the ring itself, but more often it will not have a chance to show its displeasure because another disastrous situation will already have developed, with the lubricating oil. Sometimes, if a relatively high ash-content or inadequately de-gummed oil is used, the ring will be glued solidly in its groove by varnish and carbonized oils. More often, the temperatures prevailing in that section of the piston skirt adjacent to the exhaust port will cause a breakdown of the oil film in that area and the piston will seize. And this can happen even though a generous margin of safety still exists all around the rest of the piston skirt. A common, and highly sensible solution to this problem is to move the combustion chamber pocket away from the bore axis, toward the back (inlet) side of the cylinder. This measure shrouds more of the piston crown's exhaust side under the squish band - which becomes crescent-shaped, instead of being a symmetrical ring - and reduces heat input there from combustion (the skirt will still be getting plenty of heat when the exhaust port opens) enough to provide a more even distribution of heat around the piston skirt. Then, with piston-skirt temperatures evened-out, a slightly higher compression ratio may be used without incurring seizure, or localized overheating of the piston ring.

There is another solution to the problem that has nothing whatever to do with the cylinder head: you simply add metal to the piston crown, and that, too, will tend to equalize skirt temperatures - but it also makes the piston heavier. Even so, it is a solution much-loved by manufacturers, as adding thickness in the piston costs virtually nothing, while any departure from symmetry in combustion chamber configuration entails multiple machining operations (it being extremely difficult to cast, with sufficient accuracy, the combustion chamber's small volume) and machining-time is expensive.

There may be another reason for employing an asymmetrical combustion chamber, and/or relocating the spark plug from its normal position over the bore axis. In loop-scavenged two-stroke engines, the fresh charge is directed upward, and at, the rear cylinder wall, as it emerges from the transfer ports. Ideally, the mixture streams converge and sweep up and over at the top of the cylinder to clear away exhaust products and push them out the exhaust port, following the rear cylinder wall upward, and then curling back smoothly under the cylinder head. In practice, the scavenging stream tends to be much less ordered in its habits, and the general turbulence can make it leap and dodge all over the place, impinging strongly at one point and only eddying at others. This leads, in some engines, to a reshaping and repositioning of the combustion pocket - the purpose of such changes being to aid scavenging by using the combustion chamber's form to give the scavenging stream direction.

In such cases, the spark plug may also be moved to a position where it will be washed by the mixture stream, which tends to cool the plug between firings, and thus make the engine somewhat less sensitive to plug heat range. Also, as noted before, the plug may be moved away from the combustion chamber center to create a slightly longer path for flame travel, which lowers the rate at which pressure in the cylinder rises during the combustion process and, in some instances, makes for smoother running. To a lesser extent, the same treatment may be used to combat a tendency toward detonation, as the lower pressure-rise rate gives all the pockets of end-gases time to lose their heat into the surrounding metal. This last effect is, of course, better obtained with a conical combustion chamber, rather than by offsetting the plug. Incidentally, moving the spark plug over too close to any edge of the bore is usually poor practice: At times, particularly when starting from cold, the piston ring will scrape oil off the cylinder walls and pitch it up at the cylinder head, and if you place the spark plug in the line of fire, it definitely will show a weakness for oil-fouling.

HEAD / CYLINDER SEALING

A major problem with cylinder heads on high-output engines that began life as low output engines is persistent leakage around the head/cylinder joint. The combined increases in temperature and pressure seem always to be too much for the joint, and you will find evidence of fire puffing past on the surfaces after disassembly even though you may not have observed anything out of the ordinary when the engine was running. This leaking will occur even if you have retained the engine's stock compression ratio, and it may become very serious if the head has been thinned to get a compression ratio increase. Many manufacturers, perhaps most, feel some awful compulsion to skimp on section thicknesses when they make a cylinder head, a habit that often stands revealed as a questionable economy when you test their handiwork on a dynamometer: first, the thin sections often do not have the cross-sectional area required to transfer heat away from the head's lower surface quickly enough to keep the spark plug temperatures stabilized; second, most of these cylinder heads are secured to their cylinders by only four widely-spaced bolts, which presumes heavily on their beam-strength to maintain a tight seal at the joint.

This last situation becomes especially marginal when metal has been machined away to raise the engine's compression ratio, and the stock head gasket (usually cut from light-gauge, soft aluminum) will in many cases not be strong enough to hold even the pressure increases involved in a simple switching of exhaust systems. Shave the head (which both weakens the head's beam strength and increases the forces acting upon it) and you'll very likely find that it becomes impossible to hold the head/cylinder seal- the gasket will fail after only minutes of running. Also, attempting to use the stock cylinder head, in either standard or modified form, often will increase the heat input around the spark plug to such extent that the engine becomes impossibly fussy about plug heat range. Use a plug cold enough to avoid trouble at maximum output, and it will foul at anything less than full-throttle operation. There is nothing like masses of metal to equalize the temperature gradients through the cylinder head, and - sad to say - those masses are not provided in many stock cylinder heads.

Cylinder head design also can strongly affect overall cylinder cooling. When the cylinder head's lower surface is cooler than the cylinder itself, heat will be drawn away from the latter; conversely, a cylinder head can also put heat into the cylinder if the situation is reversed. All things considered, the engine's best interests probably are served by isolating, to such extent as is possible, the cylinder and head- which means restricting the contact area at the cylinder/head joint to a narrow sealing band which bulges to encompass the hold-down bolts, or studs. In that way any cooling problems will be isolated, and can be dealt with separately. That, of course, assumes that it will be possible to improve cylinder cooling should such improvement become necessary. Actually, making a new cylinder head is fairly easy (it can be either cast or simply machined from a block of aluminum) while the cylinder itself presents a far more difficult problem in fabrication. So you may very well want to use an oversized, deeply-finned cylinder head to help cool a particular engine's stock, cast-iron cylinder. And if that should be the case, remember that you'll need a maximum contact area between head and barrel, and surfaces that will seal without any kind of gasket. There is a very sharp temperature gradient across any joint, and even a solid copper gasket presents one more pair of surfaces across which heat must flow.

You may find that providing a seal between the head and barrel is one of the more difficult facets of the overall job. As I have said, stock aluminum gaskets are almost certain to fail, being a bit weak at ambient temperatures anyway -and impossibly frail at the temperatures to which they will be subjected. Copper is a better material, for while it is nearly as soft as aluminum at ambient, its hot-strength properties are better. Copper is soft enough to make a good gasket in the annealed state, but hardens in use, and must be re-annealed frequently to keep it soft and thus retain its properties as a gasket. Brass should never be used as a gasket material, but steel may be used if it is very thin and has one or more corrugations rolled, in rings, around the bore - in the manner of the head

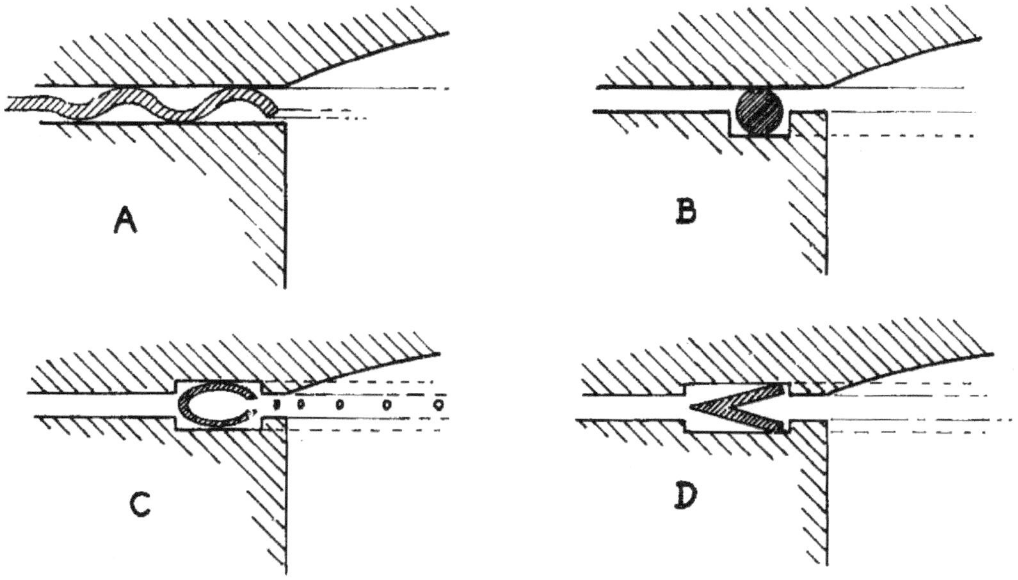

Shown here is a rolled-steel head gasket (A); a soft copper wire located in a groove (B); a hollow, gas-pressurized O-ring (C); and a V-sectoin stainless sealing ring also sealed by gas pressure (D).

gaskets used in some automobile engines. You can also get a good seal by machining a narrow groove in the cylinder's upper face and inserting in it a soft copper ring (made from wire) to bear against the head's lower surface. Other, even better seals may be had with gas-filled metal O-rings, piston rings (they'll work here, too) and one of the best sealing rings I've seen has a V-shaped section, laid on its side, with the V's point aimed away from the bore. Gas pressure tries to force the V open, bringing one arm to seal down against the cylinder while the other is pressed against the cylinder head. Another sealing ring that works in roughly the same fashion is a hollow metal O-ring with vent-holes drilled through from its inner diameter to admit gas pressure from the cylinder - which expands it outward and thus creates a seal even between somewhat uneven surfaces.

Nominal compression ratios, as I have said before, have little meaning in high-output two-stroke engines. However, you can work with trapped compression ratios almost as effectively as by measuring cranking pressures. An engine's trapped compression ratio is the ratio between the cylinder volume at the moment of the exhaust port's closing and the volume with the piston at the top of its stroke. To find this, you must first measure the combustion chamber volume, with the piston in position at top center. The job can be done with the engine assembled, using n graduated cylinder and pouring in oil until the level comes up to the spark plug hole. Or you can calculate the volume. When the combustion chamber has a simple shape (part-spherical, conical or cylindrical) I prefer to do the job by calculation, but more complex shapes send me scurrying for a can of oil and a graduated cylinder. In fact, the process of actual measurement may appeal to you as a regular thing, because you will need a graduated cylinder for more than this single task, and a slide-rule may not be a part of your basic equipment. In any case, remember when figuring the compression ratio, that it is not the ratio between piston displacement and combustion chamber volume, but between cylinder volumes from the point of exhaust port closing to top center, as in the following formula :

$$CR = \frac{(V_1 + V_2)}{V_2}$$

Where CR is compression ratio
V_1 is cylinder volume at exhaust closing
V_2 is combustion chamber volume

Traditionally, compression ratios have been measured "full stroke". That is to say, V, would represent the combustion chamber volume plus piston displacement from bottom center to tog center. Thus, a combustion chamber volume of 28cc and a piston displacement of 250cc, calculated full-stroke, would be

$$CR = \frac{(250 + 28)}{28}$$

CR= 9.93:1

But a far more realistic figure is obtained when V_1 represents the cylinder volume above the upper edge of the exhaust port, and if we assume that our hypothetical engine has an exhaust port height equal to 45-percent of stroke, then V_1 becomes 55-percent of piston displacement plus V_2, and calculation goes like this:

$$CR = \frac{((0.55 \text{ x } 250) + 28)}{28}$$

CR= 5.91:1

Coincidentally, that compression ratio (5.91:l) is very nearly all a non-squish combustion chamber will permit in an otherwise fully-developed two-stroke engine. With small-bore engines you may push the compression ratio up to perhaps 6.5:l without serious consequences, using a non-squish cylinder head, but that is very near the limit. Good squish-band cylinder heads, on the other hand, permit compression ratios up to as much ns 9.5:l in motocross engines with exhaust systems that provide a wide boost without any substantial peaks, but for road racing engines I cannot recommend anything above 8.5:1 even when unit cylinder size is only 125cc. You will find that higher compression ratios than those suggested can produce marvelously impressive flash readings on a dynamometer; as soon as the engine has a chance to get up to full temperature, the output will drop well below that sustained by an otherwise identical engine with a lower compression ratio. Sustained, and not flash horsepower, is what wins races.

EXPANSION CHAMBERS

Walter Kaaden was chief engineer of MZ's racing department through that firm's glory years on the Grand Prix circuit, and in that capacity Kaaden advanced the state of the art with regard to expansion chamber design very considerably. And one day while discussing the subject he remarked, only in jest, "You'll know when you have the design right, because the chamber will then be impossible to fit on the motorcycle without having it drag the ground, burn the rider's leg, or force the relocation of one or more major components". Of course, all present had a fine laugh, but the joke contained a large and bitter kernel of truth. In point of fact, that odd, bulky bit of exhaust plumbing we call an "expansion chamber" (a poor term for the device, but widely used) is exceedingly difficult to accommodate neatly on a motorcycle. Routed underneath, it is an acute embarrassment in terms of ground clearance even on a road racing machine and fights a losing battle with rocks on an off-road bike. Curled back along the motorcycle's side, it can force changes in the position of fuel tanks and frame tubes - and always roasts the rider's leg and/or forces him to ride bow-legged. Just as bad, it fiendishly assaults the ears of everyone for several hundred yards in every direction. and has done more to make the motorcycle - and the man astride one – unpopular than all the Wild Ones movies, and tabloid headlines of One-Percenter's misdeeds, put together.

Attended as it is by these manifold inconveniences, one almost (but not quite) wonders why we bother with the expansion chamber. Unfortunately, damnable nuisance that it unquestionably is, there is nothing else in the engineer's bag of tricks that comes anywhere close to matching the boost a two-stroke engine gets from a properly designed expansion chamber exhaust system. For that reason, it has become the ubiquitous helpmate of the high output two-stroke engine, and for that reason it will be with us until we all change over to electric motors or gas turbines. And until that time, experimenters will be tossing away stock mufflers and trying different expansion chambers as a major part of their endless quest for ever-higher performance.

Actual percentage improvements between engines fitted with their standard mufflers and the same engines with expansion chambers will vary greatly. A lot depends on how good (or bad) their muffler happened to be, and on carburetor size, porting, etc. - any of which can impose limits that cannot be totally compensated by even the best of expansion chambers. In most cases, however, the improvement will be in the order of 10- to 25-percent, and when the expansion chamber is given a bit of help from changes in timing, and the rest, it then becomes possible to get improvements ranging from 50- to (in some instances) over 100-percent. This difference is widely appreciated, even by those who know absolutely nothing about the expansion chamber itself and have no direct experience with the device, and that accounts for the brisk sale of accessory chambers as replacements for stock exhaust systems. It also has led many an enthusiast to construct an expansion chamber of his own design.

Sadly, the real result of most people's shade-tree experimental work is simply to discover that it remains possible to bring down on one's head all of the expansion chamber's considerable disadvantages without being compensated by an increase in performance. Or, as I heard one experimenter comment, looking bemused at the chamber he had cobbled together for his motorcycle, "It doesn't make much power… but it sure is noisy." He was being funny, but I didn't laugh, because the only thing that distinguished him from his fellows was that he was honest about the results; most of the others do no better - but aren't willing to admit that they have made a big mistake.

Where does everyone go wrong? Usually, it is the result of simple, uncomplicated ignorance regarding the inner workings of the expansion chamber, which-all the folklore surrounding the device notwithstanding-are absurdly uncomplicated. Using a mixture of sonic wave behavior and controlled backpressure, the expansion chamber helps pull exhaust gases out of the cylinder during the initial parts of the exhaust/transfer process and hauls the fresh charge into place- and then reverses itself to prevent the charge from escaping out the exhaust port. To illustrate the point, let's watch (in slow motion) the activity through a single operating cycle, from the time the exhaust port opens and through the transfer phase until the exhaust is once again closed. From beginning to end, the process takes only about 3- to 4-thousandths of a second.

THE BASIC PROCESS

When the exhaust port cracks open, gases still under a considerable pressure burst out into the exhaust tract, forming a wave front that moves away at high speed down the port and headed for less confined quarters. After traveling a comparatively short distance, this wave reaches the first part of the expansion chamber proper-which is a diffuser (commonly called a megaphone). The diffuser's walls diverge outward, and the wave reacts almost as though it had reached the end of the system and is, in the manner of waves explained in the first chapter of this text, reflected back up the pipe toward the cylinder with its sign inverted. In other words, what had been a positive pressure wave inverts, to become a negative pressure wave. The big difference between the action of the diffuser and the open end of a tube is that the former returns a much stronger and more prolonged wave; it is a much more efficient converter (or inverter) of wave energy.

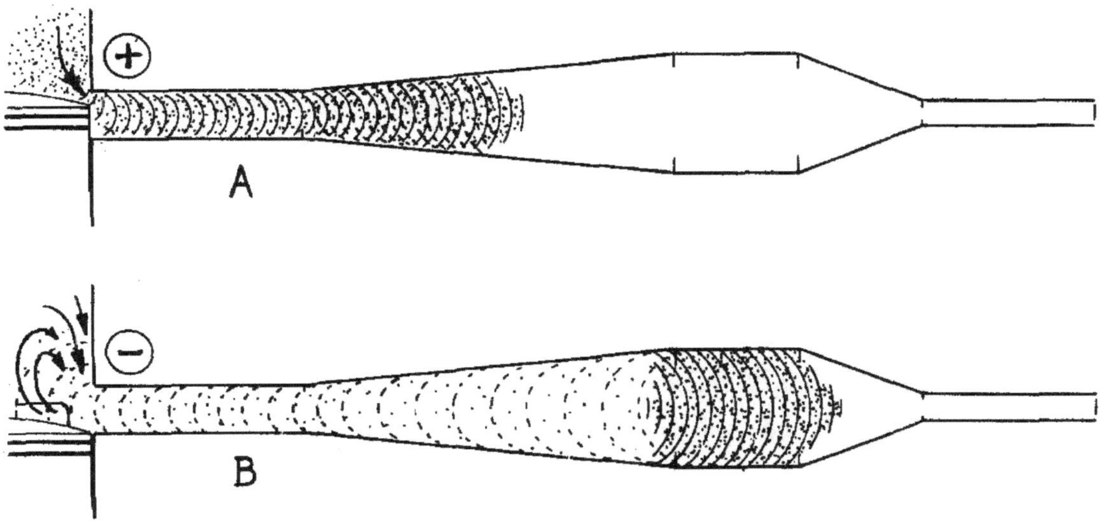

When the exhaust port opens, a strong positive pressure wave is formed and quickly moves into the diffuser (A) where part of its energy is reflected back as a negative pressure (B) to help clear the cylinder and draw the fresh charge up through the transfer ports.

As the initial wave moved down the diffuser, the process of inversion continues apace, and a negative pressure wave of substantial amplitude and duration is returned. Also, overlaid on this is the effect of inertia on the fast-moving exhaust gases, and the total effect is to create a vacuum back at the exhaust port. This vacuum is very much stronger than one might suppose, reaching a value of something like minus-7 psi at its peak. Add that to the plus-7 psi (approximately) pressure in the crankcase working to force the fresh charge up through the transfer ports and you will better understand how the transfer operation is accomplished in such a very short time. Obviously, too, this combined pressure differential of almost one atmosphere is very helpful in sweeping from the cylinder the exhaust residue from the previous power stroke. It's all a lot like having a supercharger bolted on over at the engine's intake side - but without the mechanical complication.

Years ago, the exhaust system ended right behind the diffuser. That was the arrangement on the old supercharged DKWs, and we saw stub megaphones used on the Greeves scramblers of the fairly recent past. Those devices did a job in clearing exhaust gases from the cylinder, and helped the fresh charge up from their crankcase, but their vacuuming effect was very much a mixed blessing: their problem was that they didn't know when to stop vacuuming, and would pull a sizable portion of the fresh charge right out of the cylinder. Horsepower being more or less a direct function of the air/fuel mass trapped in the cylinder at the onset of the compression stroke, this aspect of the pure megaphone's behavior was highly undesirable, and the two-stroke engine was not to come into its own in racing (where power is vitally important) until after a cure was found for the problem.

Here, our original wave reaches that "cure". Following the diffuser, and after perhaps a couple of inches of straight-walled chamber, the wave encounters a converging cone that effectively constitutes a closed end to the expansion chamber. A part of the wave energy will already have been inverted by the diffuser and sent back to the cylinder, but there is enough of its original strength left to rebound quite strongly from that closed end, and it reflects with its original, positive, sign. In due course of time, this wave arrives back at the exhaust port itself, stalling the outflow of the fresh charge. Indeed, it will momentarily reverse the flow there, stuffing what might otherwise have been lost back into the cylinder. The net result of all this activity on the part of the expansion chamber - first pulling and then pushing at the fresh charge to hold it in the cylinder - is a big boost in power. In fact, it is the only thing you can do to a two-stroke engine that will

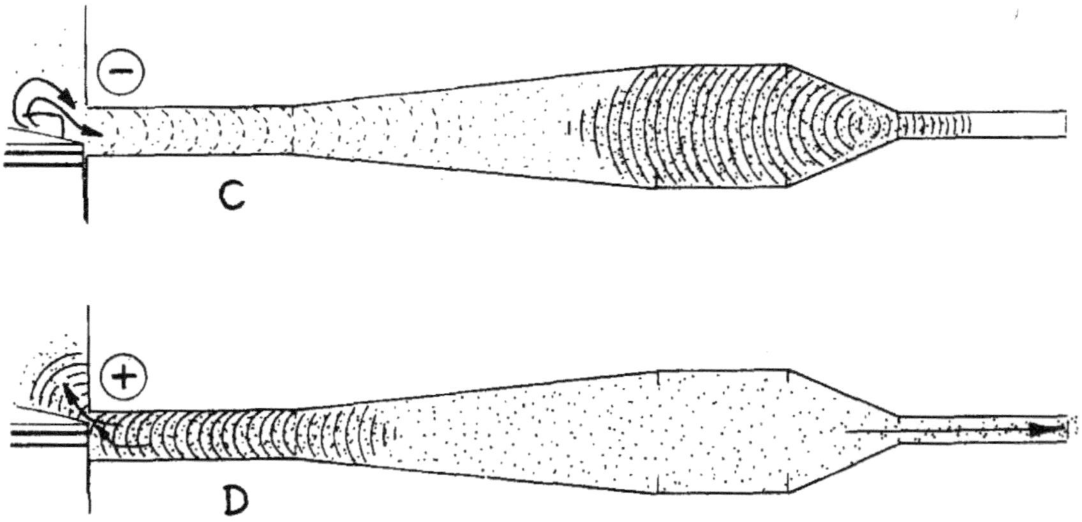

The vacuum created by the diffuser is strong enough to pull some of the fresh charge out into the pipe (C) but the original positive wave reflects back from the chamber's closed end to recharge the cylinder as the port closes (D). Then pressue in the chamber bleeds down *via* the outlet pipe before the next cycle begins.

clearly be felt in the seat of your pants; you don't need a dynamometer to find the difference.

As was mentioned before, the expansion chamber is not purely a sonic wave device: Back at the closed end of the chamber there is an outlet pipe, and it is too small to keep the pressures inside the chamber equalized with atmospheric pressure. Consequently, there is an abrupt pressure rise inside the chamber, toward the end of its operating cycle, which is felt at the engine's exhaust port and plays a very large part in preventing charge loss.

This entire process can work wonderfully well - and it also can fail miserably if the various elements of the expansion chamber are not properly dimensioned. All of the various waves and pressure sucking and surging about the exhaust port must operate in agreement with the engine's requirements. When they disagree, the result is worse than can be obtained at a much lower price paid in time and money with the stock muffler. As it happens, the motions of those waves are stubbornly tied to exhaust gas temperature, and supremely indifferent to what the engine would prefer in terms of their arrivals. The time intervals between the initial wave departure, and the return of its reflected components is a function of wave speed, and the system's lengths. Thus, as wave speed is subject only to the laws of physics and exists as something one must simply use without altering, the task of designing an expansion chamber for some particular application is to establish lengths, diameters and tapers that will use the pulsations within the exhaust system to the engine's benefit.

TUNED LENGTH

We may start by determining the proper length through the entire system back to the expansion chamber's closed end. That task requires that we know the speed at which sonic waves travel within the chamber, and therein lies a great difficulty. As noted previously, these waves' velocities are determined largely by the temperature of the gases through which they are propagated - and that factor, temperature, various continuously in the course of a single operating cycle. Exhaust gases emerge from the cylinder at about 1200° F and have very nearly (about 800° F) the same temperature back in the outlet pipe. But expansion within the chamber itself cools them (prior to recompression and reheating back in the baffle cone) to perhaps 500° F., or less, in the midsection, and a wave docs not move as rapidly through those cooler gases. It is possible to calculate fairly exactly the temperatures at all points throughout the system, but that is a very complex thermodynamic problem and certainly beyond the capabilities of the layman. Indeed, honesty compels me to admit that it is not a problem I would like to face without a computer and the assistance of someone experienced in that kind of work.

Happily, in this instance it is possible to arrive at a satisfactory solution to the problem by determining wave speed -by starting with the answer and working back. In short, you can measure a lot of existing expansion chambers *known* to be effective, and by comparing their lengths, exhaust port timings and the speeds at which the engines develop their power, eventually come up with a figure for wave speed representing a workable average for a whole range of high-output engines. My own research, conducted along the pragmatic lines just described, was begun in about 1960 and I arrived at a conclusion in 1964 that has required only slight modification over the succeeding eight years. That conclusion was, and is, that one may use a wave speed figure of 1700 ft/sec in combination with the anticipated engine speed at maximum power to arrive at a system length (measured between the exhaust port window and the point of mean reflection in the cone that constitutes the closed end of virtually all expansion chambers). That figure provides an excellent starting point for the system, as it represents a high average and any error will merely result in a lower-than-projected power peak. Actually, the addition of more examples to my charts in recent years make me inclined to think that something like 1670 ft,/sec is more accurate, but I still use the 1700 ft/sec figure as a starting point, and subsequently shorten the system slightly, perhaps an inch, if tests indicate that the power peak obtained with the chamber is too low.

Using that high-average figure for wave speed (or indeed any figure your fancy dictates, if your findings contradict my own) you may establish the exhaust system's tuned length by means of the following formula:

$$L_t = \frac{E_o \times V_s}{N}$$

Where L_t is the tuned length, in inches
E_o is the exhaust-open period, in degrees
V_s is wave speed, in feet per second
N is crankshaft speed, in revolutions per minute

For example, in an engine with an exhaust-open period of 180-degrees, and a power peak at 7000 rpm, and using the 1700 ft/sec figure for wave speed, then,

$$L_t = \frac{180 \times 1700}{7000}$$

$$L_t = 43.7 \text{ inches}$$

That length is, I must again stress, measured from the exhaust port window back to a point slightly more than halfway down the baffle cone at the end of the system. The exact point, and how to find it, will be dealt with shortly, along with an explanation of why we use a cone to close the system instead of a flat plate - and how the taper of that cone influences an engine's power curve. First, we'll consider the size and taper of diffusers.

DIFFUSER PROPORTIONS

Arriving at the proper size for a diffuser is, thanks to the work of several researchers, almost entirely a cut-and-dried proposition. Diffusers should have an 8-degree included taper for maximum energy recovery, and an outlet area 6.25-times that of their inlet. Thus, a diffuser to be attached to an inlet having a 1.5-inch diameter should have an outlet of 3.75-inch diameter. Outlet diameters for diffusers of all inlet diameters may be determined in the following manner:

$$D_2 = \sqrt{D_1^2 \times 6.25}$$

Where D_2 is the diffuser outlet diameter
D_1 is the diffuser inlet diameter
6.25 is the outlet/inlet ratio constant

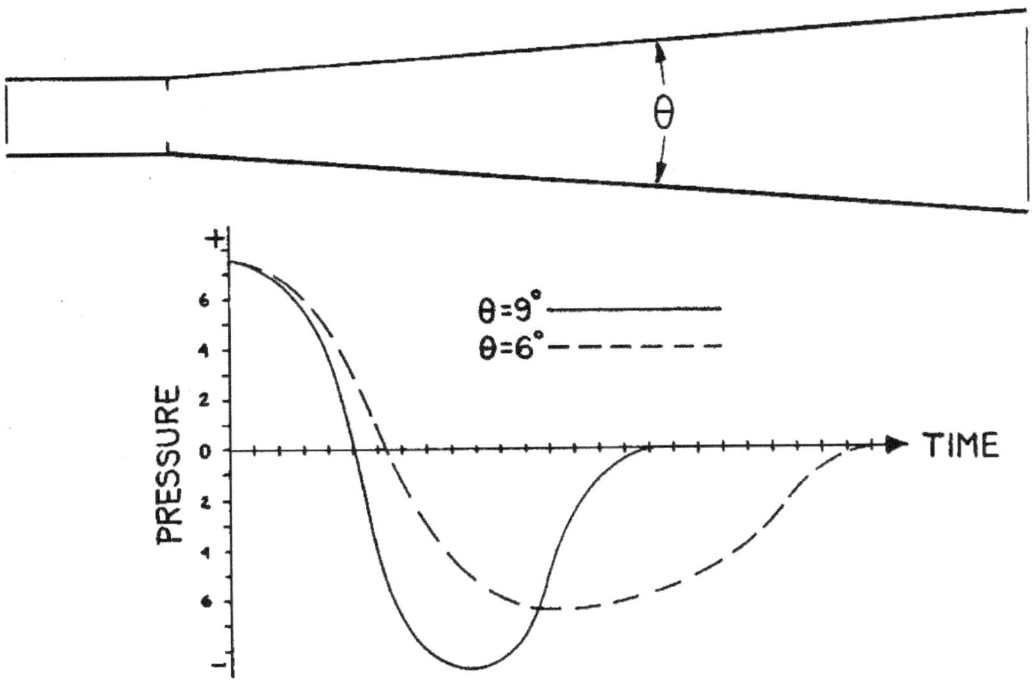

Wider angles of divergence in diffusers return stronger reflections of shorter duration and both narrow and intensify power output.

Although these diffuser diameters, tied to the 6.25 constant, remain the same, diffuser length may be varied, as there are reasons for using diffuser tapers other than 8-degrees. That taper does the best overall job of energy recovery, but it is possible to get a stronger inverted wave with diffuser tapers greater than &degrees, at the expense of wave duration. Conversely, one also may extend the wave duration by accepting some diminishing of its amplitude with shallower tapers. A long wave duration spreads an engine's power band; a short-duration wave with high amplitude is best for maximum power at peak revs. Diffusers having tapers of more than 10-degrees return a wave of such brief duration as to be almost useless even for a road racing engine coupled to the rear wheel via a multi-speed, ultra-close ratio transmission, and also are rather inefficient in terms of energy recovery. For that reason, I do not recommend that you use a diffuser taper greater than 9-degrees even when planning an expansion chamber for a road racing machine, as you may otherwise find it impossible to keep the engine operating within its power band. At the opposite extreme, do not try to use anything below a 5-degree taper diffuser in an expansion chamber for an off-road motorcycle. You will find that even a 5-degree taper results in a diffuser that is almost impossible to accommodate within the system's tuned length, and that it returns an inverted wave to the exhaust port too feeble in amplitude to be very effective in scavenging a two-stroke engine. As a rule of thumb, I would suggest employing diffusers having tapers of 8- to 9-degrees for road racing, small displacement motocross bikes (up to 250cc) should have tapers of 7- to 8-degrees, and for big motocross bikes, tapers of 6- to 7-degrees. These last usually have more horsepower than they can comfortably apply to the ground anyway, and one does not therefore give away anything by extending their power range at the expense of peak horsepower.

All of the tapers given are for single-taper diffusers. In the future we shall be seeing more of multi-taper diffusers, which I first saw on the Yamaha GP racing machines and which have since begun to appear on most of Japan's expansion chamber-equipped production motorcycles. These multi-taper diffusers would seen) to be intended to accommodate themselves to the fact that a wave, passing clown a diffuser - or any containing vessel with diverging walls -tend to separate from those walls at the diffuser entry unless the taper is slight, but is willing to accept substantially greater angles of divergence once it has moved into the diffuser far enough to have expanded and slowed. From this behavior, if I understand it correctly, one must conclude that the best diffuser would be one diverging along exponential lines -like a trumpet. Of course, such a shape would be excruciatingly difficult to fabricate, and I think it is for that reason Yamaha and Suzuki chose to employ instead diffusers with two or three tapers. My own investigation into this matter is still in its very early stages, and I cannot offer any opinion (much less

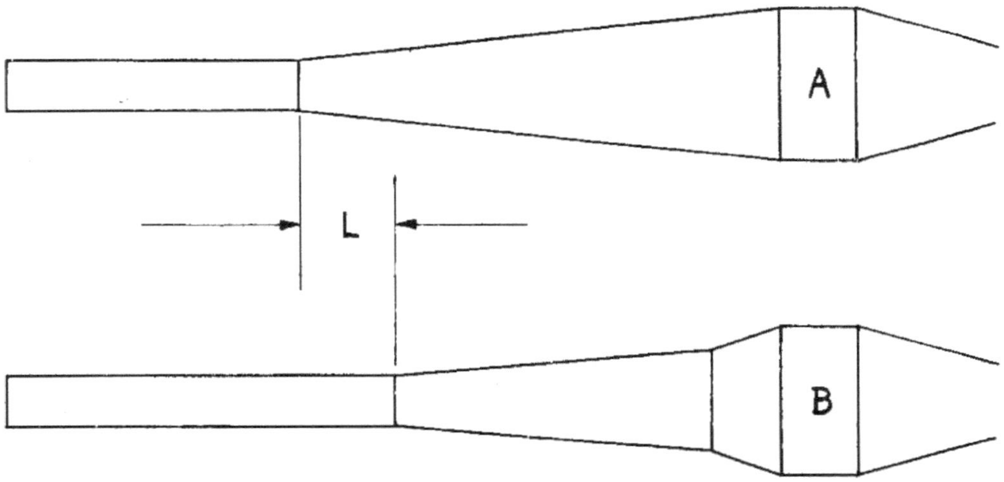

"Staged" diffusers (B) are shorter than those with a single taper (A) and make available extra length (L) to be used in the lead-in pipe to give power range or to increase volume for peak horsepower.

facts) except to note that Yamaha has been working with diffusers tapered 7-degrees for 70-percent of their length, and the remaining 30-percent tapered 12-degrees. Both Yamaha and Suzuki have recently begun to connect these two-stage diffusers to the exhaust port with a lead-in pipe that is also tapered 2- to 3-degrees. This slight taper probably is provided to minimize flow resistance and thereby enhance the lead-in pipe's ability to inertia-evacuate the cylinder, rather than to function as part of the diffuser.

Incidentally, the use of two-stage diffusers also facilitates coping with the often-conflicting requirements of expansion chamber volume and lead-in pipe length. A diffuser proportioned 7° / 12°, 70% / 30%, is somewhat shorter than a diffuser having a straight 8-degree taper when both are constructed to the same 6.25:1 outlet/inlet area rule. This reduction in diffuser length may be added at the lead-in pipe, or at the chamber's parallel-wall section, where the diffuser and baffle cone are joined. In either case, with a longer lead-in pipe or added chamber volume, the overall effect is to increase power output below the power peak - with volume having its most pronounced effect high on the engine speed scale, near the power peak; length added to the lead-in pipe brings about a somewhat more serious drop in maximum power, but also lends a marked increase in low speed torque. These effects, it should be noted, tend to modify the choice of diffuser tapers, as a short, steep-taper diffuser provides room for a longer lead-in pipe or added chamber volume - both of these tending to offset the power range-narrowing influence of such diffusers.

Some departures from the 6.25:l area ratio rule may also be desirable on occasion: when a lead-in pipe diameter is exaggerated, relative to exhaust port area and cylinder size, in the interest of low-range performance, you may find that following the ratio rule results in an expansion chamber of such monstrous diameter that it simply cannot be fitted on the motorcycle for which it is intended, and in that case an appropriate downward adjustment in diameter is indicated.

Should this consideration of inconvenient bulk lead you to depart from the dimensions required by my formulae, think twice before you succumb to the temptation to flatten the expansion chamber. In the first place, you'll upset all the area progressions through the diffuser; an 8-degree diffuser, flattened ever so slightly, is no longer an 8-degree diffuser. Moreover, even if you calculate the areas so that you have a rounded wedge with the correct inlet/outlet area proportions, wave energy recovery will still suffer. Those waves simply do not like being puffed through anything but a cone; even less do they like a cone that has been dented or notched to clear a frame tube or to provide ground clearance. They can "feel" every change in cross-section in the containing vessel. They are, however, willing to follow even the most abrupt jog in the system: you can resection the diffuser cone and kink it all over the place to make the

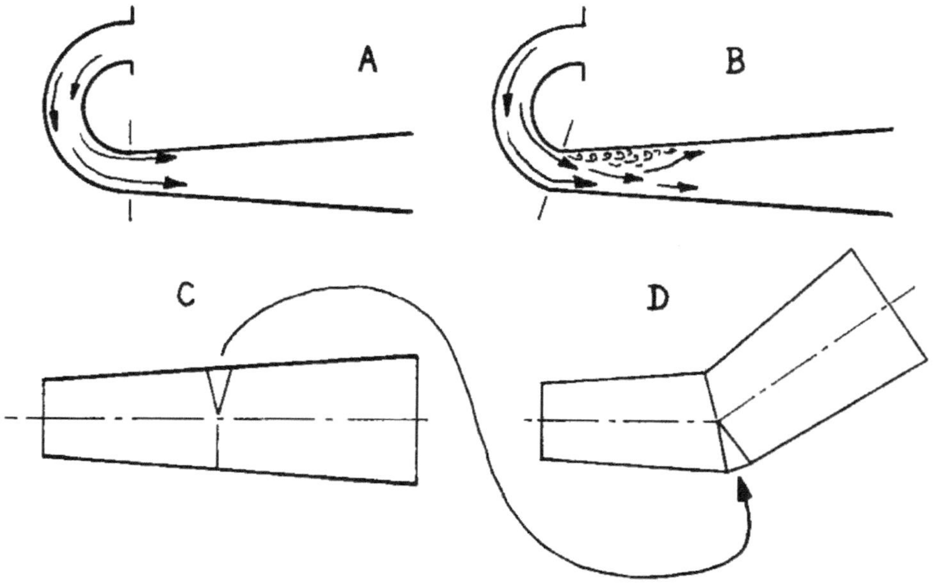

A smooth entry into the diffuser gives good flow (A) but a kink (B) creates turbulence. Sharp bends in the diffuser have little effect on power and may easily be made using the pie-slice method shown.

expansion chamber fit the motorcycle, and the wave will never know the difference. Sonic waves may be able to feel even the most minute changes in section; they will make any turn you can build into the system without slowing or losing any of their energy.

The only part of the system where you must be careful to provide smooth turns is up at the lead-in pipe and at the entry to the diffuser. Through that section, gas velocity is very high, and while the *wave* won't care about sharp jogs, such jogs will have a bad effect on gas flow - which is a different matter entirely (gas flow involves the movement of matter; a wave is just energy, and being without mass, is also without inertia and therefore cares nothing about sharp corners. At least, that is very substantially the case here, where the velocities, etc. involved are low enough to fall within the scope of Newtonian physics; Einstein's unified field work is hardly applicable at wave speeds of such limited magnitude).

If efficiency were the only consideration in current expansion chamber design, it would be possible to depart fairly substantially from the round cross-section back at the chamber's middle portion. Actually flattening the chamber is still not a good idea, but it can be squared off somewhat without greatly compromising its basic power-enhancing properties. Two years ago, I would have recommended precisely that sort of modification in instances where installation was a problem. Now, with the drive against noise well and truly underway, any departure from round is to be considered poor design practice. Why? Because those waves I have been talking about are very strong, and will make even a round-section chamber's walls ring like a bell (to be specific, a cow-bell) just like the engine is shooting marbles out its exhaust port. These pulses, which are strong enough to set up a ringing even in the relatively stiff walls of a round-section chamber, will make any flat areas in the expansion chamber's walls pant in and out like a drum-head. This vibration is of course transmitted into the surrounding atmosphere as a hellishly loud noise, and no matter how effective a muffler you may add back at the chamber's outlet pipe, the motorcycle's overall noise output will nonetheless be very high. The noise source just described can only .be minimized by either making the chamber out of very heavy steel, or by giving it a shape that resists pulsing; the round-section chamber may ring somewhat, but it cannot actually pulse in and out even when made of very light-gauge material. This pulsing of the chamber's walls has another highly undesirable side effect: it makes the permanent attachment of a bracket or heat shield very difficult. Most fasteners will fairly quickly fracture from the severe vibration, leaving the heat-shield to drop away - which is bad, but not as bad as when the same vibration fractures a major mounting bracket and the entire expansion chamber conies adrift. For all these reasons, the round-section expansion chamber, although inconveniently bulky at times, really seems to be the best choice.

BAFFLE CONES

We have already noted that the baffled end of most expansion chambers is conical. This cone lends the chamber rather more pleasing lines than it would have with a flat end, but that is not its reason for being. The reason is that if we end the chamber very abruptly, with a flat plate, the wave reflections away from it will also be very abrupt: strong, but of a duration too brief to provide the desired port-plugging effect except within extremely narrow limits in engine speed. A conical baffle, on the other hand, extends the wave reflection time (as reflection occurs down its entire length) and, because its effects are thus felt over a wider engine speed range, the engine's useful power band is broadened. Obviously, here, a long, gently tapered baffle-cone will extend an engine's power range more than a shorter, more sharply tapered cone in the customary trade-off between range and peak power. These tapers should be, in most cases, twice that of the diffuser used in the expansion chamber. Thus, in a chamber having an 8-degree diffuser, the baffle-cone should be tapered 16-degrees. That is the rule in general. However, wide variations are possible and may be employed to cope with a specific situation. The largest taper angle you should use is 20-degrees; the smallest, 14-degrees. And you may, to obtain a particular effect, "mismatch" diffusers and baffle cones in any combination.

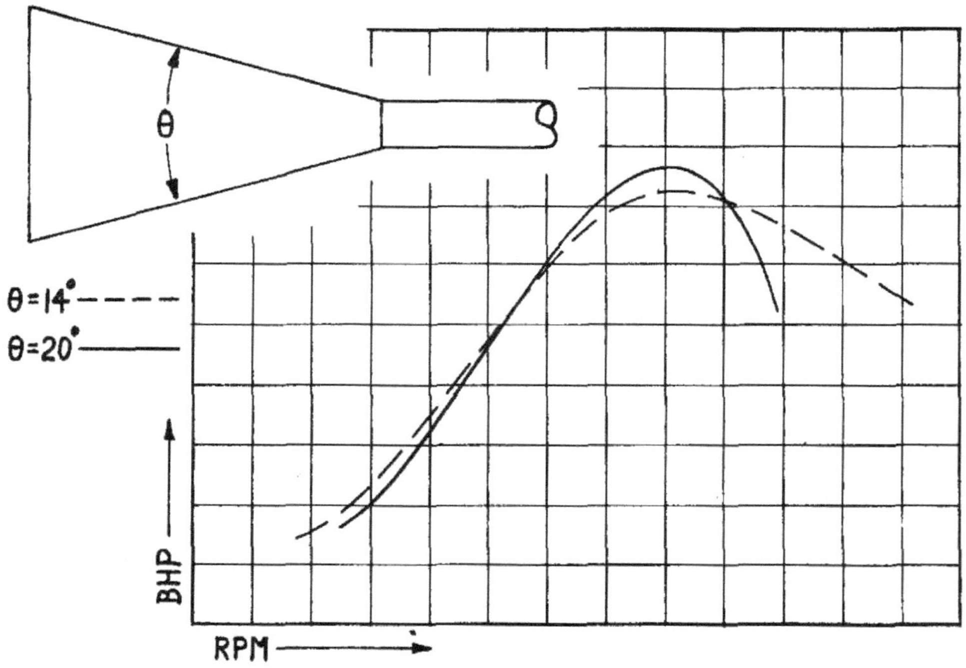

Changes in baffle-cone angle primarily influence the shape of the power curve past the point at whish maximum power is obtained.

The thing to remember is that there is a peculiar side to the power-range broadening effect of the baffle-cone: most of it is on the part of the engine's power curve *past the horsepower peak.* Thus, for an engine that has proven to be rather fragile when pressed beyond its rpm red-line, you may terminate the expansion chamber with a 20-degree baffle cone, and rest assured that if the system's tuned length is established to place the horsepower peak, say, 500 rpm below the danger mark, the engine will resist very strongly any effort to get it spinning faster.

My experience in dynamometer-testing various expansion chamber configurations has shown that a 20-degree baffle-cone gives a good, strong power peak - and then simply cuts the engine dead, in terms of output, if you try to force the revs any higher. A 15-degree baffle-cone, in contrast, gives a somewhat lower maximum output - but helps the engine maintain its output well after the peaking speed has been exceeded. The implications of this influence on an engine's power curve should be obvious: motocross bikes can gain in engine flexibility from a long, gently tapered baffle-cone, but if you are "tuning" for some young man who often forgets to protect the engine from over-exerting itself by changing gears, then you can use a 20-degree baffle cone. It will remind him about the gear lever by chopping the power drastically every time he tries to use too many revs.

There is another thing you need to know about those baffle-cones before you can design your own expansion chambers: they do, as stated earlier, reflect over their full

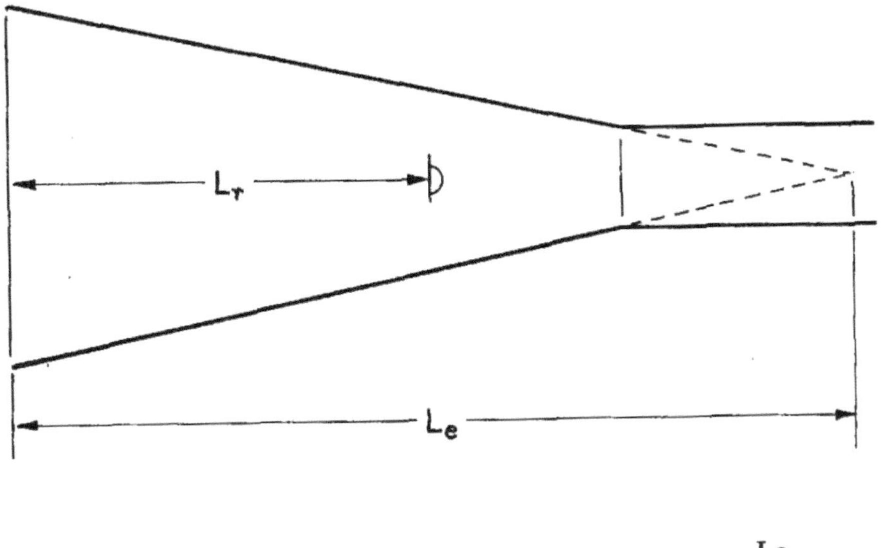

The mean point of reflection inside the bafflle-cone is $Lr = \frac{Le}{2}$

length any wave entering them, but there is not an *even* reflection, for reasons too numerous and involved for serious discussion here. What you do find, however, is that there is a "mean" point of reflection which is, as I said earlier, slightly more than half-way down the baffle-cone's length. The "tuned length" we discussed before, the length so critical to the expansion chamber's performance, is actually the distance from the exhaust port window, at the piston face, measured along the exhaust system's center-line out to this point of mean reflection. This midway point seems to be at the halfway point of the *complete* cone, which would be half the length of the cone if the cone were complete, right out to a sharp tip, instead of being truncated at its small end to make room for the outlet pipe. You can find this point of mean reflection either by drawing the complete cone, measuring, and dividing its length in half, or by using this simple formula:

$$L_r = \frac{\left(\frac{D_2}{2}\right) \times \cot(A_2)}{2}$$

Where L_r is the distance from the baffle-cone's inlet to the mean point of reflection
D, is the baffle-cone's inlet diameter
A, is half the baffle-cone's angle of convergence (i.e. for a 16-degree cone use 8-degrees, etc.)

While we are on this particular subject, I will also give you the formula for finding the length of a cone, given its taper, major and minor diameters. The formula is as follows :

$$L = \frac{D_2 - D_1}{2} \times \cot(A)$$

Where L is length
D_2 is the cone's major diameter
D_1, is the cone's minor diameter
A is half the angle of divergence, or convergence.

OUTLET PIPES

Having gotten past the business of diffusers and baffle-cones, we can proceed onward to the lengths and diameters of lead-in and outlet pipes. The latter will, if the rest of the expansion chamber is proportioned fairly closely in accord with the advice I have given, have a diameter between .58- and .62-times that of the lead-in pipe, and a length equal to 12 of its own diameters. It is simply a pressure-bleed resister, which prevents the free escape of exhaust gases from inside the expansion chamber and thereby creates a backpressure to enhance the port-plugging efforts of the wave reflected by the baffle-cone.

This outlet pipe is much more sensitive to diameter than length. The rule I have provided for diameter will get you very close to the optimum, if the rest of the chamber is proportioned correctly (or at least as I have indicated, which I hope will amount to the same thing) and the length, of 12-diameters, will also be very close. Even so, they will only be "close" and I have found few instances, in my own experimenting, when a session of cut-and-try on the dynamometer did not lead me to a slightly different diameter and length for best results. You probably do not have a dynamometer, and that bit of news is in all likelihood not something you wanted to hear. Nonetheless, it is true. You will not get optimum results here simply by following the instructions I provide. In fact, a tad of adjusting is required, to meet a particular set of requirements, to make any strictly-formula expansion chamber give maximum performance. But there is a very particular problem associated with the outlet pipe that you should know about: Simply stated, the problem is that if you make the outlet pipe too small in diameter, or give it too much length, (both tending to over-restrict the chamber's outlet passage in terms of flow capacity) then there will be a price to be paid in overheating. I frankly do not know the mechanism involved in this; I only know that if the outlet is too restricted, engine temperature is very greatly elevated - and Yamaha's Naito has said that the same unfortunate result will be observed if the expansion chamber's baffle-cone is tapered too abruptly. From what I have seen, this increase in temperature is especially sharp at the piston crown, which tends to be the power-limiting part in a two-stroke engine in any case. So, you should keep a sharp eye on your engine's piston. There will be temperatures high enough to darken the underside of the piston crown, due to baked-on oil forming there, when the engine is healthy; just don't ignore the warning sign you see when that oil begins to char. The next thing you see, after you see ash, will be a hole.

Actually, the worst side of this whole matter of selecting the right outlet pipe is that while power rises quite sharply as you work your way down from a too-large outlet pipe diameter, there will be only a slight decrease in power output when you have gone too far in restricting the outlet. Therefore, unless you have a dynamometer and instrumentation for measuring engine temperatures, make all errors on the large side of the diameter you get by multiplying lead-in pipe diameter by the 0.58-0.62 factor I have provided. In other words, if the factor-derived diameter (taken from a lead-in pipe diameter of say, 1.75-inches) is from 1.015 to 1.085-inches, then you should not try to use a piece of one-inch tubing for that outlet pipe, as its nominal diameter is an outside measurement. Subtract for a wall thickness of, say, 0.049-inch, and you'll have a passage only 0.902-inch in diameter. That isn't large enough to be safe - nor would a tube having a nominal diameter of $1\ {}^{1}/_{8}$-inches, with an inside diameter of 1.027-inches be entirely safe. When skirting that close, go up to the next largest available diameter - in this case it would be $1\ {}^{1}/_{4}$-inches –and run the engine long enough and hard enough to permit a valid "reading" of the piston crown's underside. If it shows no sign of excessive heat, you can try a slightly smaller outlet pipe, and then check the piston again.

LEAD-IN PIPES

There probably is a better, closer, rule for determining outlet pipe diameter -if only because the rule I have offered is tied to the diameter of the expansion chamber's lead-in pipe, which is very, *very* difficult to determine on a strictly theoretical basis. For engines having exhaust port timings and port widths typical in racing terms, the expansion chamber's lead-in pipe should have a diameter providing an area 10- to 15-percent greater than that of the port window. But that only applies, I have found, when the parameters are as stated, and only then when a power curve very distinctly biased toward maximum output at maximum revs is desired. For motocross applications, the diameter chosen will be one that can be used in combination with a considerable pipe length to broaden the power range. In some instances, lead-in pipes are used with cross-sectional areas representing 150-percent of the exhaust window area. As a general rule, you may assume that the manufacturer of your particular engine knows more about lead-in pipe areas than you, or I, and you can't get into trouble following his lead. You may, by ignoring the possibilities in other directions, miss an optimum by some smallish percentage - but you won't get into trouble.

The thought may have occurred that my instructions are, in this regard, something far short of precise. And so they are, for good reason: the choice of lead-in pipe diameter must be shaped not only by unit cylinder displacement, port timing/area, and according to the application you have in mind for the engine-but also with an eye toward the lead-in length, and the configuration of the diffuser to which the pipe attaches. All of these things have their effect, but I have not as yet been able to isolate each item well enough to arrive at a quantitative pattern. So, for the moment, I make do with a qualitative understanding and a couple of handy rules-of-thumb: the first you already know (regarding the rough relationship between the port and pipe cross-section area); the second rule (more a suggestion, really) is that for maximum horsepower only, you should give the lead-in pipe a length equal to from 6 to 8 of its diameters, while for a broader power curve (and at some expense to the maximum) you may want to use up to 11-times pipe diameter. All these lengthy are, of course, not just that of the pipes themselves, but also include the distance from the pipe mounting flange through the port to the piston face. A final note on lead-in pipe length: if, for reasons of easier installation, or a broadening of the power range, or both, you decide upon a lead-in pipe length greater than 8-times its diameter, plan to increase its diameter slightly. Resistance to flow is increased with length, and this can and should be offset by giving the gases a larger passage. Ideally, this reduction in resistance would be accomplished by using a lead-in pipe having a slightly-diverging taper (2 or 3 degrees) but that may represent a more difficult fabrication job than you care to undertake.

After you have finished the job of designing an expansion chamber exhaust system, the thing will have to be translated from a set of dimensions into metal, or you'll never know whether or not the design is any good. You can, of course, build the whole thing yourself, but I would not recommend that approach. Especially, I would caution against attempting to roll your own cones, as this is the kind of job that takes endless hours when you don't have the proper tools and is likely to turn out badly anyway. Go to a sheet-metal shop and pay them to make your cones; they'll get it done in much less time and if they're sharp you'll get cones having the right lengths, tapers and diameters for very little more money than you would spend just to buy the materials. Bent tubing, to be used in making the lead-in pipe, can be obtained in a variety of diameters and bend-radii at most muffler shops. Welding? Unless you have the equipment and are competent at welding light-gauge metal, that too should be hired from a professional, for the finished item can get to look very scabby given the best of workmanship. As for tolerances, the only thing you really must watch is that the fractions of an inch do not add up to upset the distance from the exhaust port window back to the baffle cone. An inch there, will move the power peak about 500 rpm. The same increment in length, up at the lead-in pipe, produces a slight reshaping of the entire power curve. Through most of the system, try to work to within $^{1}/_{16}$-inch tolerances.

When testing the finished item (to recap some things already noted), remember that changing the taper of the baffle-cone changes that Dart of the - power curve past the power peak, while changes in the lead-in pipe length mostly influence the low-revs part of the curve. Increases in volume are effective mostly in adding area under the power curve right at the power peak; reducing outlet pipe diameter will, if the existing pipe is too large, boost power over the entire range, but requires that you keep a close watch over piston-crown temperature. There are other, more subtle things that may need correcting: for example, those oh-so-helpful sonic waves are not always content just to reverberate up and down the entire length of the expansion chamber. As it happens, each part of the system tends to develop its own individual resonances, and the lead-in pipe, diffuser, baffle-cone and outlet pipe will each have their own little wave system rattling back and forth, with other resonances of lower frequencies occurring in paired parts of the chamber. In most cases, these incidental wave patterns go unnoticed, but sometimes they will fall into mutually-reinforcing activity at certain engine speeds and combine to force humps and hollows into what would otherwise be a nice, even power curve. When these do occur, often a slight shifting on lengths is enough to keep them from marching locked-step and interfering with the job you are trying to do with the primary wave motions.

Far more common, when you fit an expansion chamber exhaust system on some engine intended for an unassuming muffler, is that the balance between the mixture delivered by the carburetor and the requirements imposed by conditions in the combustion chamber is destroyed. Regrettably, the high output two-stroke engine is to a very large extent gasoline-cooled, requiring rather more fuel mixed with its air than is needed to satisfy the simple chemical combination of oxygen and hydrocarbons in the burning process. In light of this, you will appreciate that an expansion chamber arranged to provide an enormous boost at some point along the engine speed range makes it absolutely impossible to jet the carburetor for clean running at all engine speeds. The way it works is that an engine's torque peak sets the mixture-strength requirement - with regard to cooling the piston crown below its melting point, squelching detonation, etc. - and if that mixture is too rich for lower, or higher, engine speeds, then the power range suffers. Sometimes, jetting for that torque peak will make the mixture so rich that the engine will hardly run at all in its low and middle ranges. There will be more on this subject in the chapters on carburetion and cylinder head design; it should be enough to say, here, that for truly broad range performance, it may be necessary to compromise in terms of expansion-chamber effectiveness to stay within the tolerances permitted by the mixture an ordinary carburetor will deliver. You can do that, and then take up some of the slack in engine performance with a higher compression ratio than you would dare use with an expansion chamber that gave a big narrow-range boost.

Two years ago, you wouldn't have been bothered with anything beyond the expansion chamber's outlet pipe; today, you'll have to fit a muffler, and it is important that you know what such an end-fitting means in over-all performance. Properly managed, the addition of a muffler to your expansion chamber will have absolutely no effect on power output, and will increase only slightly the exhaust system's size and weight. Trying to accomplish the same thing with a racing four-stroke engine would be difficult to the point of impossibility, but the two-stroke's expansion chamber must have a restricted outlet in any case, and it cares little whether the restriction is provided by a simple bit of pipe, or by a muffler. I have dyno-tested a number of expansion chambers, made as replacements for stock mufflers, that actually showed a gain in power with an add-on muffler. Of course, these chambers clearly had too-large outlet pipes, and the same result could have been obtained with pieces of smaller-diameter tubing. Still, it would be the *same* result, which means that there is no performance penalty to be paid when an expansion chamber's outlet pipe is capped with a muffler.

The only real problems you will encounter in muffling an expansion chamber exhaust system are: muffling effectively; and muffling without adding too much flow resistance. The latter problem is of minor importance if you have access to a dynamometer, but assumes critical proportions when you must rely strictly on seat-of-the-pants testing. Adding too much restriction, with the muffler, will produce precisely the same effect as when the outlet pipe is made too small: drastic overheating of the engine. Predicting flow-resistance through a muffler is very much more difficult than for a straight pipe, so the trick is to use a muffler that offers a straight-through passage for the exhaust gases, and a passage of such diameter that it will add little or no resistance to that already provided by the chamber's outlet pipe. With a muffler meeting that specification, the expansion chamber simply will not know that it isn't releasing its pulses straight into the atmosphere. There will be no special problem in finding that kind of muffler, as several manufacturers - their interest stimulated by the recent AMA rule requiring that all motorcycles in sportsman-class racing be muffled - are now producing straight-through silencers to spigot-mount on outlet pipes of various diameters.

This brings us to the problem of *effective* silencing, and I take no pleasure in telling you that none of the mufflers being offered for sale as this is written are particularly effective. They are, of course, better than nothing, but they will not reduce the noise output from your motorcycle down to even the 92 dbA maximum specified by the AMA's rule. Worse, most of them rely on fiberglass packing for their sound-damping properties - and this fiberglass fairly quickly disintegrates and is blown out with the exhaust gases, or loads with carbon and oil (which kills its ability to attenuate sound waves), or both. Only frequent replacement of the fiberglass packing will keep such mufflers in working order, and we may very shortly face federal regulations that specifically forbid the use of fiberglass in mufflers. And the same regulations probably will specify a maximum noise level for even off-road vehicles somewhat below the AMA's 92 dbA limit - a limit that is not now being met by fiberglass-packed mufflers in perfect working order.

These several difficulties should convince anyone that a different approach to the problem of effectively silencing the expansion chamber is required. Lacking a more effective solution to the problem, we may eventually be forced to revert to a straightforward muffler in place of the expansion chamber and live with the loss of power and performance that entails. I do not believe that will be necessary, as I stumbled upon a phenomenon a few years ago that meant very little at the time but now assumes major importance: The then-existing general racing regulations required that a motorcycle's exhaust system terminate at some point forward of the rearmost edge of the back tire, and I was planning to race a bike with its cylinders reversed to provide rear-facing exhaust ports (for reasons that were important, but not pertinent here). The only major flaw in this scheme was that even with the motorcycle built on a longish wheelbase and with its engine located well forward, there was not quite enough room for the exhaust pipes within the length allowed by the rules. The expansion chambers themselves would fit, but there was some 12-inches of outlet pipe left hanging back behind the rear tire, and not enough room to curl these outlet pipes back within the limit. While groping for a solution, I hit upon the idea of simply sliding them forward, inside the baffle cones. There, they would still function as pressure-bleed resisters, and further contemplation led me to the conclusion that the expansion chambers might even work better with their outlet pipes positioned inside. With the forward end of the outlet pipe located at the chamber's maximum diameter, ahead of the baffle-cone, there should be a somewhat stronger reflection from the baffle, and that might very well give the engine a somewhat better boost. Or so I thought.

Anyway, I gave the scheme a try, and while certain other modifications prevented drawing any absolute conclusions from the experiment, the bike did prove to be very fast, and it seemed certain that while my "inside stingers" might not offer any real power advantage, they probably were at least as effective as those attached in the more conventional manner. But that is not to say that I did not notice a difference - and the difference was in sound. With those inside outlet pipes, the typical expansion chamber crackle was very noticeably subdued. That made sense, as the chambers' outlet to atmosphere was taken from a point where the pulse was at its lowest amplitude - rather than from the high-pressure area at the tip of the baffle-cone.

More recently, I have been able to perform a series of tests using a dynamometer, to see if my inside-stinger idea (which, I had learned, was an idea also advanced by an obscure German researcher some several years earlier) had any real merit. An expansion chamber was constructed with its baffle-cone terminating in a clamp, to allow rapid changes in outlet pipes. and we tried pipes of different diameters and lengths, and moved them back and forth in the clamp to find the change in output as the outlet's forward end was advanced up the baffle-cone. The results were most interesting: There was absolutely no difference in power output with the outlet pipe in full-forward and full-back positions, but we did find a quite noticeable drop in power with the pipe's forward end pushed up to the approximately-halfway point in the baffle-cone. We also found that the same outlet pipe diameter produced optimum results no matter what the location, but that the system was rather less sensitive to outlet pipe length when the pipe's forward end was located an inch or so ahead of the baffle-cone's forward end. Finally, we found that the noise output with the forward-located outlet pipe was very much reduced: to about the same level as a conventional expansion chamber fitted with a bolt-on, fiberglass-packed muffler. I was, of course, a bit disappointed that my inside-stinger pipe did not show a big advantage in power over the conventional variety (there may be a slight broadening of the power curve, but the differences observed were too small to offer conclusive proof). Still, by that time, the sound-damping properties of the inside-stinger arrangement had become extremely important, as they could be used in conjunction with a low-resistance muffler, located farther downstream, to meet the AMA's new noise-limit rule without any loss of performance. For that reason, this idea - the creature, indirectly, of the AMA competition rules book - did prove to be a success, even if not in precisely the form I had anticipated.

Given the disadvantages of the fiberglass-packed muffler, better designs are needed and already are beginning to appear. Yamaha, for example, have quite an effective muffler for their expansion chamber-equipped motocross motorcycles. This one consists of a perforated tube passing through a canister, with the center of the tube plugged to force the exhaust pulses out through the perforations in the first half of the tubes and, into the canister, where they escape back through the holes in the tube's second half and then off into the atmosphere. Passage through the holes, which have diameters of about 5mm, breaks up the pulse, and it is further attenuated by expansion inside the canister. My only concern here is that Yamaha's new muffler may over-restrict the expansion chamber outlet, but given that company's thorough approach to engineering and testing, that seems a remote possibility. However -and this is not my concern, but the AMA's -I doubt that Yamaha's muffled motocrosser is really quiet enough to meet the AMA's 92 dbA limit. Perhaps so; perhaps not. In any case, the expansion chamber is here to stay, and so is the movement to restrict noise. The problems of effective muffling will be solved, and I think my inside-stinger may help with the solution.

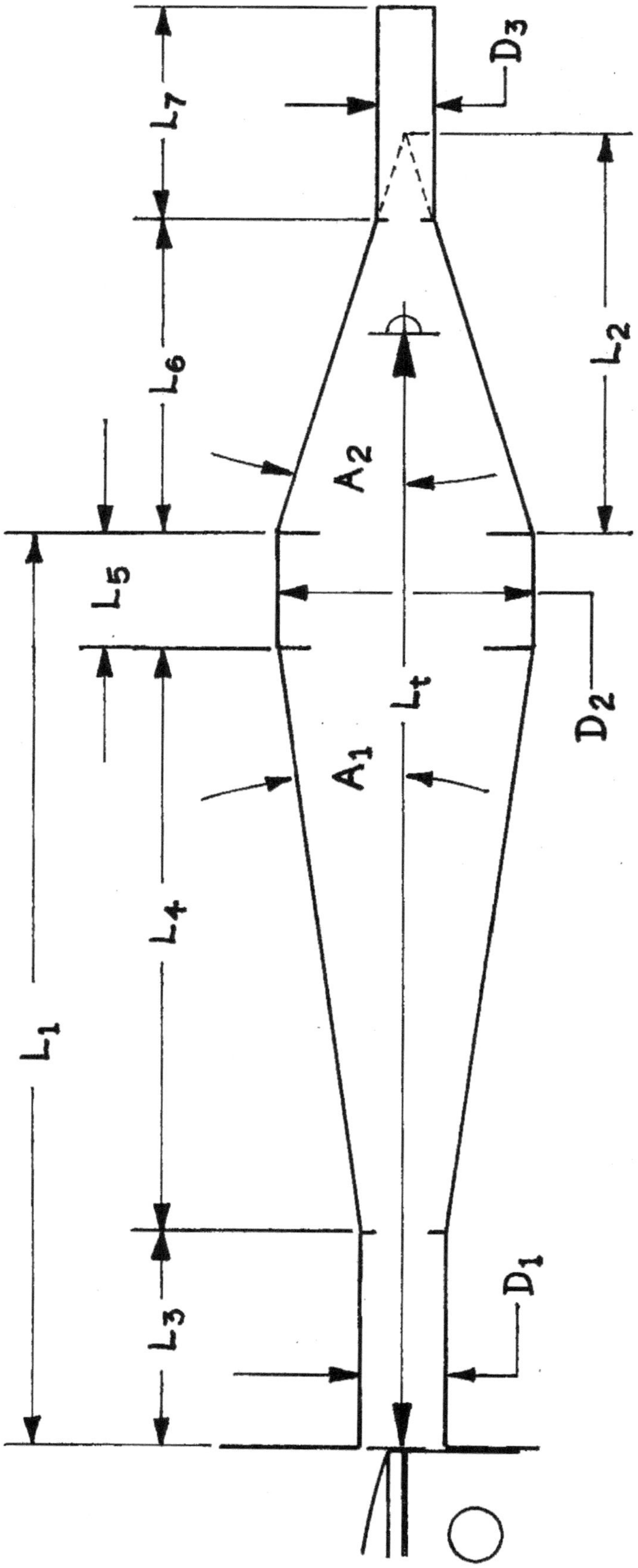
L_7
D_3
L_2
L_6
A_2
L_5
D_2
L_t
A_1
L_4
L_1
L_3
D_1

EXPANSION CHAMBERS

EXPANSION CHAMBER DESIGN FORMULAE

$$L_t = \frac{E_o \times Vs}{N}$$

Where: L_t = tuned length, in inches

E_o = exhaust-open period, in degrees

V_s = sonic wave speed (use 1700 feet per second)

N = crankshaft speed, in revolutions per minute

$$L_1 = L_t - \left(\frac{L_2}{2}\right)$$

$$L_2 = \left(\frac{D_2}{2}\right) \times \text{cotangent}(A_2)$$

$$L_3 = D_1 \times (6 \text{ to } 11, \text{see text})$$

$$L_4 = \left(\frac{D_2 - D_1}{2}\right) \times \text{cotangent}(A_1)$$

$$L_5 = L_1 - (L_3 + L_4)$$

$$L_6 = \left(\frac{D_2 - D_3}{2}\right) \times \text{cotangent}(A_2)$$

$$L_7 = D_3 \times 12$$

D_1 = (see text)

$$D_2 = \sqrt{6.25 \times D_1^2}$$

$D_3 = D_1$ x (0.57 to 0.62, see text)

A_1 = (half the diffuser's angle of divergence)

A_2 = (half the baffle-cone's angle of convergence)

USEFUL COTANGENTS:

cot 3.0° = 19.0811
cot 3.5° = 16.3499
cot 4.0° = 14.3007
cot 4.5° = 12.7062
cot 5.0° = 11.4301
cot 5.5° = 10.3854
cot 6.0° = 9.51436
cot 6.5° = 8.77689
cot 7.0° = 8.14435
cot 7.5° = 7.59575
cot 8.0° = 7.11537
cot 8.5° = 6.69116
cot 9.0° = 6.31375
cot 9.5° = 5.97576
cot 10° = 5.67128
cot 11° = 5.14455
cot 12° = 4.70463
cot 13° = 4.33148
cot 14° = 4.01078
cot 15° = 3.73205

CENTERLINE LENGTH OF CURVED PIPE

L = R x 0.01745 x angle of bend

PORT TIMING

Most of the literature dealing with the two-stroke engine's port timings makes a serious and fundamental error in concentrating on timing to the exclusion of everything else. In many instances very impressive-looking graphs will be presented, to illustrate how exhaust port timing, for example, will advance with increases in engine speed, etc., etc., but all seem to slip quietly past a couple of facts that should arrest everyone's attention: First, it should be clear to anyone who has made even a cursory study of the various two-stroke engines' port timings that no close correlation between these timings and engine speed exists. Second, it should be equally clear that the basic function of these ports - which is to permit a volume of gases under pressure to escape confinement - is not solely dependent upon time. Time is required, obviously, but the process is one that also is sensitive to area. In other words, at any given pressure differential the time required to move a volume of gases through a port in the side of a cylinder wall will be mostly a function of the port's cross-sectional area. Thus, it is not possible to treat with timing alone when planning the porting in a two-stroke engine; you must find the combination of time and area that will permit complete passage of the gas volume at hand.

This time-area requirement has been known to me a lot longer than I have been able to effectively quantify the concept. It is one thing to know that time-area numbers applicable to the whole range of two-stroke motorcycle engines probably exist, and quite another to find them. What I needed first was a mathematical expression that would combine the time factor with port area, and include an adjustment for the ratio between port area and unit cylinder volume. But while this might appear to be fairly simple, there is a serious difficulty in arriving at a value for port area. Just measuring the total port area would not suffice, for the port is fully open only for that infinitesimally brief moment when the piston rests at the end of its stroke; at all other time, during the period when a port is nominally open, some part of the port window is masked by the piston. Worse, the degree of masking varies to some extent according to differences in ratio between the on-centers length of connecting rod and stroke, and also with the absolute port-open duration. In consequence, any study of port time-area would have to be based on numbers not readily available unless one has endless hours to devote to collecting information either locked away inside the engines themselves or inside manufacturer's archives. It is possible to work out time-area numbers on a purely theoretical basis, but this approach is even more time-consuming, involving as it does some really nasty work with integral calculus equations. Also, this approach virtually demands that one make some fairly shaky assumptions in a number of areas -and that a computer be at hand if all the calculations are to be completed in this century.

All this changed when I acquired an SAE paper presented by Yamaha's Naitoh and Nomura. In this paper, time-area values suitable for two-stroke motorcycle engines are given, and while there was no explanation of how these were derived - particularly with reference to the determination of port area - they did provide the key that unlocked what had been a great mystery. While puzzling over the Naitoh-Nomura time-area values, I recalled that one of the pioneers in the field, Schweitzer, had based his calculations on a "mean" port area that was simply the aperture presented when the piston was halfway toward the fully-open position at the end of its stroke. That is to say, halfway in terms of degrees of crank-angle, and not where half the port is uncovered. For example, with an exhaust port that opens 90-degrees before bottom center, the mean port area would be taken with the piston 45-degrees of crank-angle from bottom center -which in most engines will uncover about 70-percent of the total port-window area. Assuming that Naitoh and Nomura were working along those lines, I worked out time-area values for a couple of Yamaha racing engines for which I had all the applicable data, and they fell neatly into place. Actually, the Naitoh-Nomura paper was rather obscure on this point and I was forced to do a lot of assuming, but it did provide a pattern in time-area relationships that, once unraveled and checked against concrete examples provided by Yamaha's engines, enabled me to present what I consider to be valid numbers.

SPECIFIC TIME-AREA

The numbers given here express time and area, and the ratio between port-window area and cylinder volume. They represent narrowly defined guidelines for intake, transfer and exhaust port time areas as follows:

For piston-controlled intake ports, 0.00014 to 0.00016 sec-cm^2/cm^3

For transfer ports..................., 0.00008 to 0.00010 sec-cm^2/cm^3

For exhaust ports..................., 0.00014 to 0.00015 sec-cm^2/cm^3

For rotary-disc intake valves,,,,,,,,, 0.00018 to 0.00019 sec-cm^2/cm^3

Some of you may not quite understand that expression, "sec-cm^2/cm^3". It may look a trifle intimidating, but it merely indicates that the number preceding it was derived by dividing cylinder volume, in cubic centimeters (cm^3), into the mean area in square centimeters (cm^2) of the port in question, and then multiplying the resulting number by the total time, in seconds, during which the port is open. It says, in short, "time-area per unit displacement", and it removes all the mystery that has so long surrounded port timing.

To work any time-area problem, you must first be able to convert an engine's timing, in degrees, into actual time at some given engine speed. Let us begin with a hypothetical exhaust port timing of 170-degrees duration, and an engine speed of 7000 rpm - the speed at which we intend that the engine should develop maximum power. To convert these known factors into time, we use the following formula:

$$T = \frac{\theta}{N \times 6}$$

Where, T is time, in seconds
N is engine speed, in rpm
θ is the port-open period, in degrees

Thus, $$T = \frac{170}{7000 \times 6}$$

$$T = 0.004 \text{ seconds}$$

Determining T, or time, is quite simple; the process of finding a port's "mean" area is somewhat more complicated and infinitely more time-consuming. There are two basic approaches to finding mean port area: The most direct is to simply bolt a degree-wheel to the end of the crankshaft, align it for TDC, then find the point at which the port begins to open and, finally, crank the engine over to a point halfway between the point of port-opening and bottom center (or top center, if the intake port is being checked). With the piston at that halfway mark, measure the area of the port window that is exposed and you'll have the mean port area - which represents the average size of the port-window aperture through the whole period from opening to closing.

The second method for finding mean port area, and the one I prefer, is to transfer all of the pertinent dimensions to a sheet of fine-grid graph paper and work from that. Begin by drawing a vertical line to represent the cylinder axis and then, toward the bottom of that line, add a circle representing the path followed by the crankpin. With that done, measure up from the top of the circle a distance equal to the on-centers length of the connecting rod and add a horizontal line at that point. Repeat this, measuring from the bottom of the circle, and you have created the space swept by the piston. The ports are then drawn in, flanking the line that represents the bore axis, with the exhaust and transfer ports down against the bottom of the space and the intake port situated up at the top. Finally, mark the mean port-open points on the circle, using a protractor, and connect these points and the bore-axis line with lines of precisely the length of the connecting rod, center to center. You will find that the connecting points, on the bore-axis line, fall about 70-percent down on the exhaust port, about 75-percent down the transfer ports (of which only one side need be shown) and about 65-percent up on the intake port. Draw horizontal lines through the port windows at these points, and you'll then be able to measure the mean open areas... assuming that you have accurately reproduced all the dimensions.

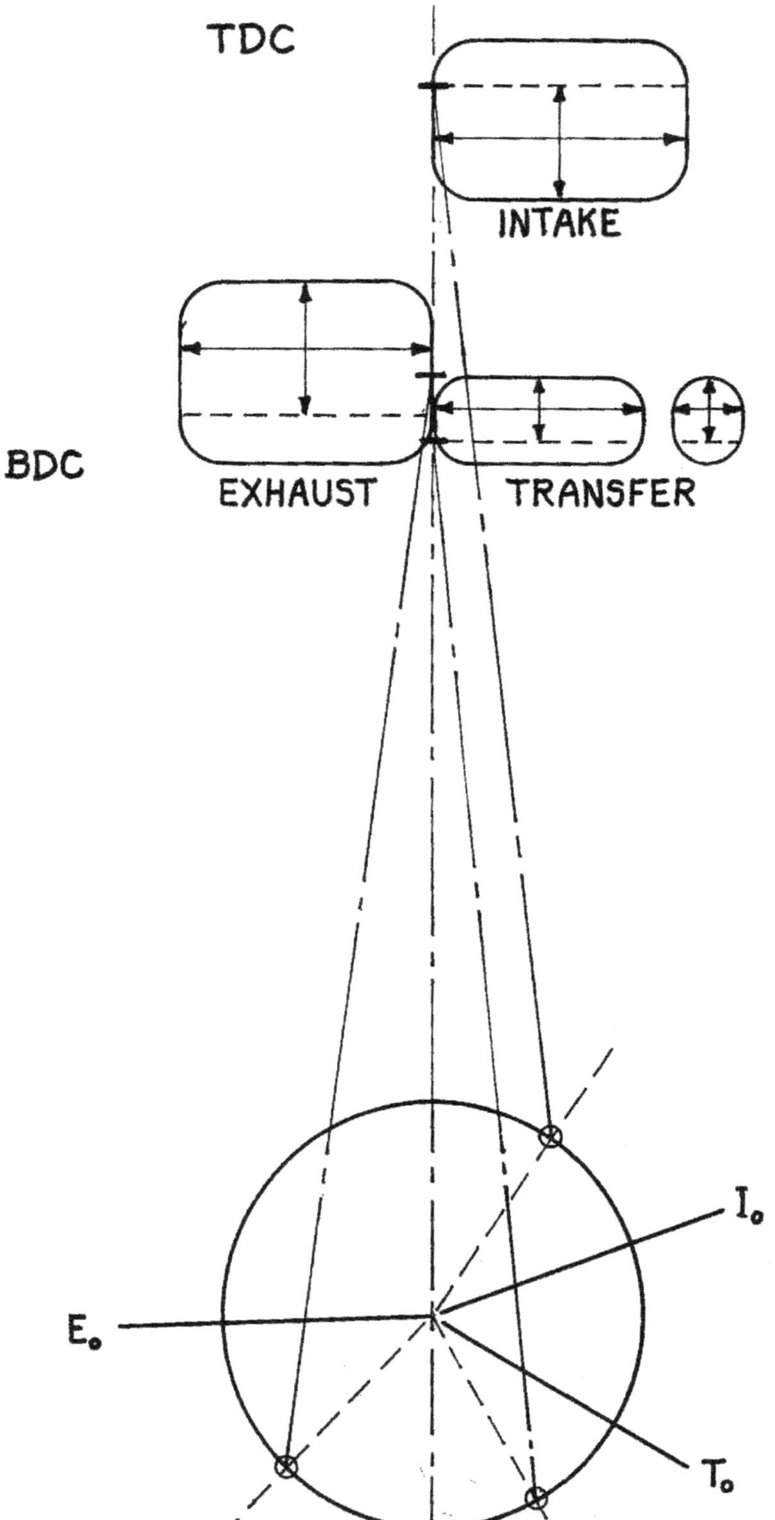

Time-area studies may be made directly from an engine, but a better method is to transfer the appropriate crank, rod, and port dimensions to a sheet of graph paper, then take mean port areas from the drawing.

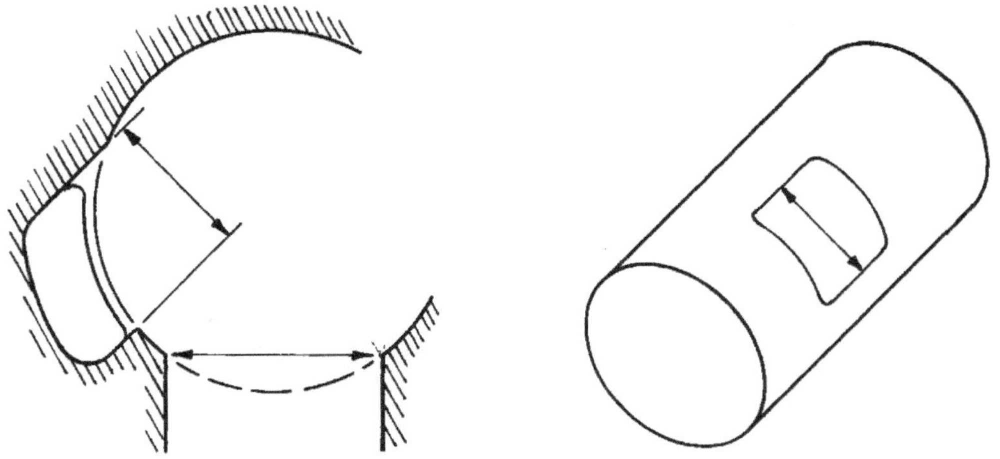

Measuer port widths across the chord unless, as is sometimes the case with transfer ports, entry is sharply angled. Angled entryways should be measured for their true cross-sectional area.

You will find that the most difficult job is to accurately reproduce the port window shapes. Exhaust ports tend to be more nearly round than square in high-output engines with wide, un-bridged ports; intake ports have rounded sides even when they are basically rectangular; transfer ports often have \ angled upper edges; and in addition to these difficulties, you must determine accurately the radii existing at the corners of all ports. So, finding the area of a given port aperture is not simply a matter of multiplying width times height. My own method for resolving this difficulty is to make a close-fitting paper sleeve, slide it inside the cylinder being checked, tape it in place and then, by rubbing around the edges of the port windows with the side of a pencil tip, transfer the window shapes to the paper. Please understand that these shapes *may* not then be transferred directly to the graph paper on which the bore, connecting rod, and crankshaft lines have been drawn. Why? Because in laying the sleeve flat, the width of the ports is exaggerated, and what you need is the true size of the aperture. My approach is to make this paper cylinder out of fairly stiffish stock, and after getting the port shapes marked on its inner surface to reverse its sides, creating a cylinder with the port window drawn outside where they can conveniently be measured with precise calipers and the true dimensions then transferred to my working drawing. A series of horizontal guide-lines drawn across the port shapes before re-rolling the cylinder is an aid in measuring the port's shape, but in the final analysis accuracy depends on one's willingness to be meticulous with all measurements and to be imaginative in concocting ways of recreating a port shape on paper. I have found that a draftsman's ellipse template is very useful in this kind of work, but it is only essential that you have a good protractor, a straight-edge, a compass, and a supply of fine-grid graph paper. By assigning a value of one-millimeter per line, and doing all drawing to scale, you can not only work with considerable accuracy, but the determining of window areas become much easier - you just count the number of squares and fractions of squares within the "mean" aperture to get area, and if you have drawn everything carefully, the margin of error will be too small to be of any particular significance.

Having found the ports' mean areas, you are ready to complete your calculations in finding each port's value in sec-cm^2/cm^3. For an example, we'll return to the hypothetical exhaust port for which we have a port-open duration (derived from 170-degrees and 7000 rpm) of 0.004 sec. We will further assume that this port has a mean area of 8.0 cm^2, and is in a cylinder having a displacement of 250 cm^3 (or, as it is more commonly expressed, 250cc). To arrive at the figure we want, in sec-cm^2/cm^3, it is first necessary to divide cylinder volume, 250 cm^3, into port area, 8.0 cm^2, and then multiply the results by time, in the following manner:

$$\frac{8.0}{250} = 0.032 \text{ ; } 0.032 \times 0.004 = 0.00013 \text{ sec-cm}^2/\text{cm}^3$$

And, referring back to the specific time-area figures already provided, which establish a range from 0.00014 to 0.00015, we learn that this particular hypothetical engine is time-area deficient, in the order of 10-percent. Increasing the port width, to bring the mean open area up to 9.0 cm^2, will raise the exhaust port's specific time-area to 0.000145, which is right in the middle of the specified range. Increasing this engine's exhaust-open duration will increase both time and mean area, and that's where the drawing on the graph paper becomes very handy, as it enables you to chart how much area is gained by increases in timing, and to establish what combination of port width, height and timing will yield the specific time-area need for the particular engine with which you are working. In most instances you will not be able to obtain the correct values by merely widening ports; a combination of increases in both time and area are required if your stock engine has a power peak at, say, 6000 rpm and you want to push it a thousand or so rpm higher.

ANGLE-AREA

With all the juggling of numbers entailed in finding correct combinations, it is advisable to find short-cuts, and one of these is to work with specific angle- area numbers instead of time area. One cannot simply substitute angle-area for time-area and ignore the engine speed factor, obviously, so I have provided charts from which you may work. Take, for instance, the exhaust port time-area/angle-area chart: you'll find that there are two line marking the exhaust time-area limits of 0.00014 and 0.00015 sec-cm^2/cm^3, with vertical lines marking off engine speed and horizontal lines for angle-area. The specific angle-area of our hypothetical example, taking the exhaust-open duration as 170-degrees and the cm^2/cm^3 figure as 0.032 (8.0/250), is 170 x 0.032 = 5.44 deg-cm^2/cm^3. Referring to the chart, we note that at 7000 rpm the specific angle-area requirement is from about 5.88 to 6.3 deg-cm^2/cm^3, meaning that more time and / or area will be needed if the engine is to operate effectively at 7000 rpm - and that an angle-area value in the order of 6.1 deg-cm^2/cm^3 is a good target figure. This process is, of course, intended to be repeated for all of any given engine's ports.

These time-area and angle-area numbers are enormously useful in planning the porting of any two-stroke motorcycle engine, and I would be happy to tell you that they represent the final word on the subject... but they unfortunately do not. Many other factors intrude upon this seemingly uncomplicated picture: Referring once again to the time-area/angle-area charts, you will observe that there is, in every instance, a *range* for what constitutes correct values, which reflects the fact that the optimum is influenced somewhat by factors beyond those already discussed. Sadly, it is in accounting for these that we must depart from the comfortable certitudes of mathematics, for the other influences are in the field of exhaust system design and characteristics, flow coefficients, nominal and effective port areas, crankcase compression ratios, resonance in the intake tract, and very likely a host of minor influences that will forever escape our attention. All these add up to a situation in which a certain amount of adjustment, done according to results observed in testing, will always be necessary. Calculation brings us close to the optimum, after which we work within the time-area/angle-area range according to an educated guess, and finally we test. Professional engineers do not do the job any differently.

TIME-AREA COMBINATIONS

Even though I cannot provide you with formulae, or even anything other than generalized comment, I can give you at least some indication as to what combinations of time-areas/angle-areas work best in specific instances: Exhaust system characteristics are, for example, very closely related to both exhaust port *and* transfer port time-areas. In a motocross engine, where power range is more important than maximum power, you would want an exhaust system that provides a relatively weak but extended duration resonant pulse (which means that it would be effective over a very broad engine speed range) and, relatively speaking, a low exhaust port time-area value in combination with a high transfer port time-area. In fact, if the engine in question is a single-cylinder 400, or something in that general displacement range, then you probably would select a transfer time-area at the very top of the range, and an exhaust time-area established at the bottom. That is to say, a transfer port time-area of 0.00010 sec-cm^2/cm^3 and an exhaust port time-area of 0.00014 sec-cm^2/cm^3. This combination of time-areas provide an exhaust timing that borrows minimally from the working stroke, which maximizes broad-range torque, and it gives the negative wave returning from the exhaust system a better opportunity of finding the transfer port still open –meaning that this negative, or scavenging pulse will be able to help pull the fresh charge up from the crankcase. Relatively large transfer port time-areas also give the fresh charge conditions allowing it to make its way up into the cylinder even if there is little or no assist from the exhaust system. The sole disadvantage of the condition described is that it does appear to permit a somewhat greater degree of mixing between residual exhaust gases and the incoming charge, as well as an enhanced tendency toward short-circuiting.

Road racing engines represent an entirely different situation, as they respond best, being tuned for maximum power and with power range a secondary consideration, to a maximum value for exhaust port time-area and a minimum transfer port time-area. At least, that will be the case when the engine in question is fitted with a typically road-racing expansion chamber exhaust system, which returns very strong but short-lived scavenging and plugging pulses to the cylinder. Also, for reasons that will be discussed later, road racing engines' scavenging is aided by high crankcase pressures resulting from strong, albeit narrow-band "supercharging" effects of sonic wave activity in their intake tracts. With the extractor effects of the exhaust system and the ramming effects on the intake side combining to aid cylinder charging, maximum power is obtained by upper-limit exhaust port time-areas (to make best use of the exhaust system) and lower-limit time-areas on the transfer side to minimize charge dilution and short-circuiting.

Carried to its extremes, the described combination of sharp, powerful pulses from the exhaust system and low-limit transfer port time-area can elevate, and narrow, an engine's power band to a remarkable degree. The two-cylinder, 350cc Yamaha TR3, for instance, has an exhaust port time-area value of 0.000148 sec-cm^2/cm^3 and a transfer port time-area of 0.000081 sec-cm^2/cm^3, and these numbers represent very nearly the maximum and minimum time-area values within their respective ranges. Power output from this engine must be in the order of 63-65 bhp, but the power is developed over such a narrow range that a 6-speed transmission with ultra-close ratios is required to hold it within limits. I am informed by a very good source that the latest Yamaha 250cc TD3 has an even narrower power band, and that the most skilled of riders has considerable difficulty keeping it on the power curve under actual racing conditions. My suggestion to

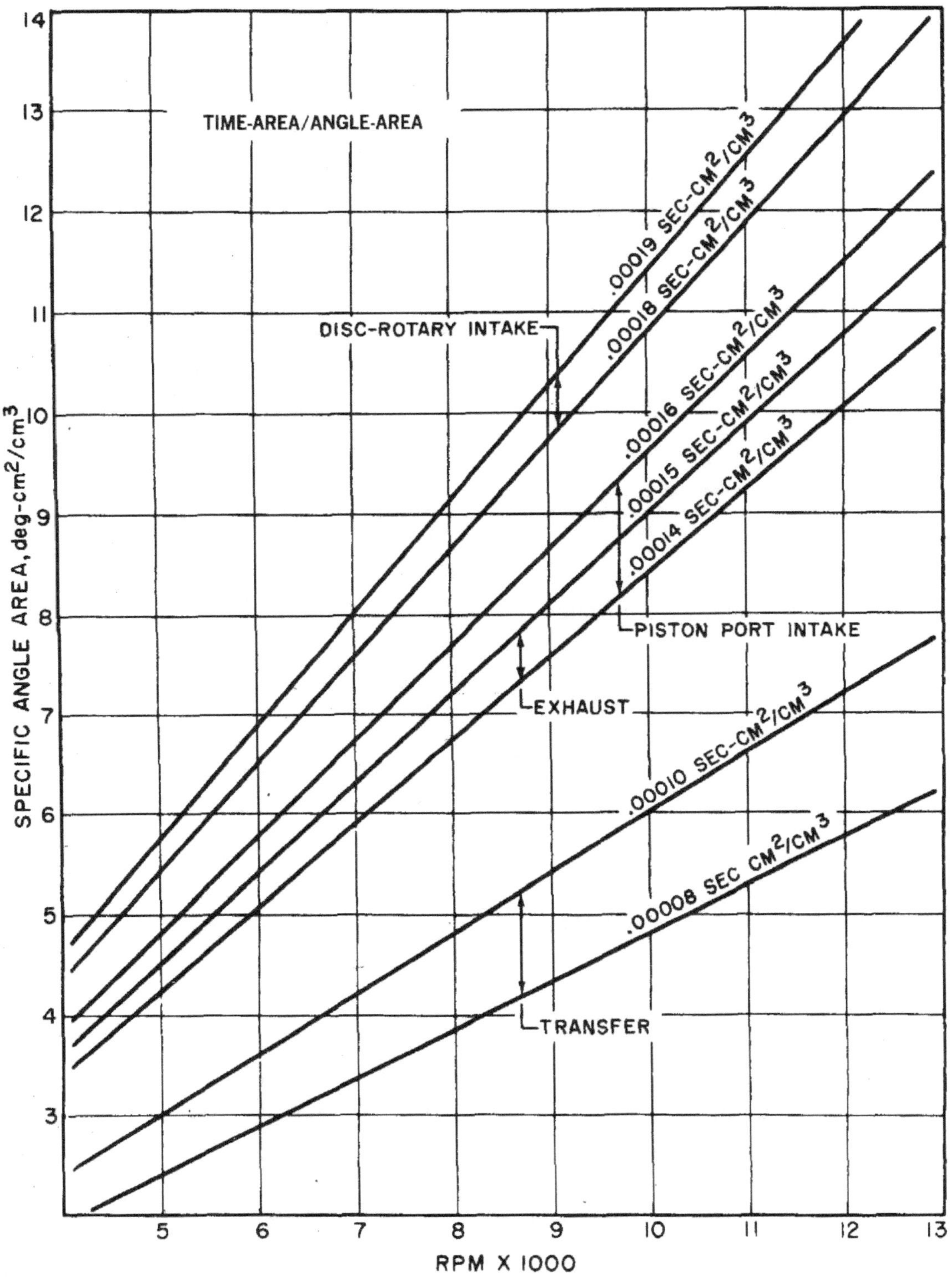

Shown here in graph form is the relationship between time-area and angle-area over a range of engine speeds.

those who have this machine, and the difficulties, is to increase the engine's transfer port time-area slightly. An increase in the transfer-open duration of only 4- to 6-degrees would probably broaden the TD3's effective power band enough to make the machine very much easier to ride, reducing maximum power by perhaps two bhp and adding about three or four bhp at the lower limit of the present range. The same applies to all two-stroke engines: increases in transfer port time-area tend to depress the power peak, but add to the power curve at lower engine speeds.

It should be understood, however, that excessive transfer port time-area, in combination with the wrong exhaust system, can lead to serious instability in running - yielding a major drop in peak power without adequate compensation in power range, and a power curve marked by humps and hollows. Thus, while engines exist in which exhaust/transfer time-area imbalances (relative to the values presented here) have not prevented quite good power outputs, such imbalances may be regarded as extreme-example anomalies more interesting for their value as curiosities then as patterns from which to work. In most engines the correct approach will be to establish time-area values that fall within the ranges suggested here, and to make adjustments within those ranges according to the conditions for which the engine is intended. Road racing engines for which 6-speed and/or close ratio gear sets are not available should be biased toward the "motocross" end of the time-area spectrum; small displacement motocross engines which commonly are coupled to fairly close-ratio, 5-speed transmissions should be biased toward the road racing specification simply because the horsepower thus gained has become a competition necessity- even though a rather peaky motocross engine is no joy to a rider.

EMPHASIS ON AREA

Taking each port individually, there is every reason to make any port as wide as possible, acquiring the necessary time-area value in this manner instead of extending the port-open duration. Reasons for moderating this approach do exist, however, in the interaction between ports and in the effects exaggerated exhaust port widths have upon pistons and rings. These reasons are discussed in detail in the chapter titled, "Scavenging", but I will I include a brief reminder here. A too-wide exhaust port will cause rings to snag and break, or wear very rapidly, and if the widening brings the exhaust port window's sides too near the transfer ports, there will be an increased tendency toward short-circuiting of the incoming charge. Obviously, excessive widening of the transfer ports can also result in ring-trapping and/or charge short-circuiting. You should also understand that widening an engine's exhaust port, increasing its time-area value without actually increasing its open duration, has much the same effect as obtaining the same increase by raising its height and thus increasing both time and area: that is to say, widening the exhaust port increases the engine speed at which maximum power is realized, while reducing low-speed power. And the same pattern is to be observed in increases to transfer port time-area, though in the opposite direction. These effects should become familiar to you, particularly as regards the exhaust port, for any increases in exhaust time-area should begin with widening the port to the maximum tolerable to the piston and rings, moving on to the business of raising the top of the exhaust port only after the limit for width has been reached. There is good reason for taking this approach, for while increases in exhaust port time-area, gained by whatever change in the port-window's shape, certainly will have the same general effect, increasing width to get more time-area has a much less narrowing effect on the power band than increases in height.

TIMING LIMITS

Width is even more important on the intake side of any engine with a piston-controlled intake port, as there are sharp limits to time-area increases gained by lengthening the port-open duration. Piston-port engines have the advantage of simplicity, but are somewhat (sometimes seriously) handicapped by the fact that their intake timing is symmetrically disposed before and after top center. There is, therefore, a strong tendency for the mixture aspirated into the crankcase during the period between intake-opening and top center to be pushed back out during the equal port-open period between top center and intake-closing. This tendency accounts for the extraordinary influence of intake-tract resonance and gas-inertia on the piston-port engine's power characteristics. The combined activities of sonic waves and the inertia of the high-velocity mixture stream can simply overpower the rising pressure in the crankcase created by the descending piston.

Ideally, intake-closing should occur at the precise moment when ramming pressure is at its peak and when that pressure is equal to the pressure inside the crankcase, as this condition will trap the greatest volume of air/fuel mixture inside the crankcase. Unhappily, this ideal can only be realized within very narrow engine speed ranges, as inertia effects diminish rapidly at lower-than-planned speeds and the natural frequency of the intake tract is determined almost solely by its (and the crankcase's) dimensions, which means that it pulses at a fixed rate, and only at one particular engine speed will it truly be working in phase with the motions of the piston. Worse, at very low engine speeds neither sonic-wave activity nor the ramming effects of gas I inertia will be strong enough to prevent the piston from displacing part of the charge aspirated into the crankcase right back out through the carburetor. All of which means that at cranking speeds, when you are trying to start an engine, the total volume of the charge being delivered into the

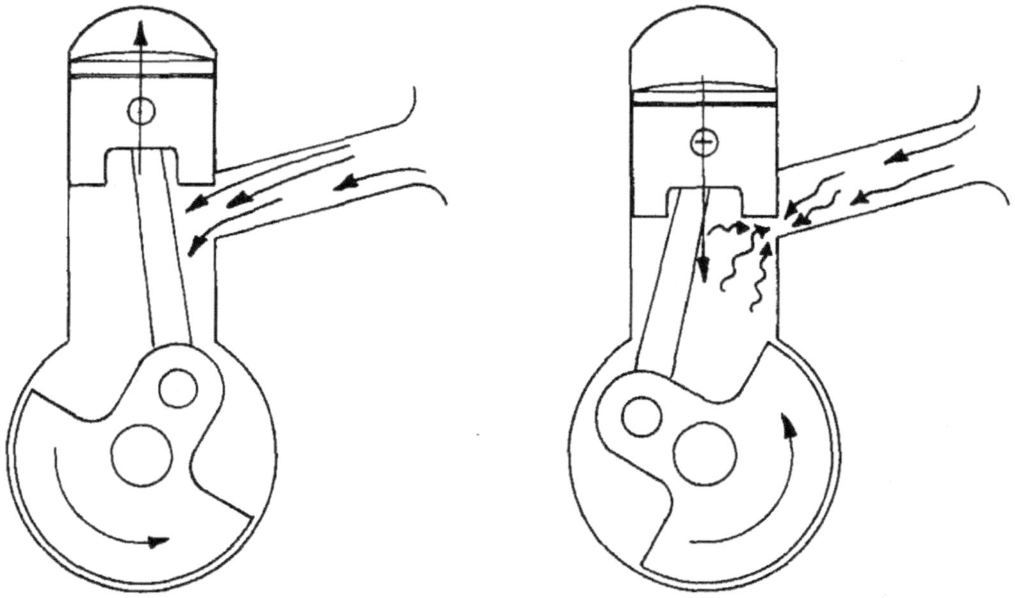

Time-area is not the only consideration in intake timing as intake closing must occur when crankcase pressure and resonant/inertia ram pressure balance.

cylinder will be determined by that which the piston displaces between the point at which the transfer ports close and the point of intake opening (which also is intake-closing). For example, in a piston-port road racing engine with transfer-closing at 115-degrees before top center and an intake port opening and closing 100-degrees before and after top center, the volume of gases actually being pumped through the crankcase, per revolution, would be only that displaced by the piston in a mere 15-degrees of crank-angle.

It is entirely possible that in the example given, starting would prove to be impossible unless the engine was actually cranked fast enough to bring it up to the point where intake-tract resonance and inertia began to have some effect. Quite obviously, this example does represent an extreme, but not one I that seriously distorts the condition being illustrated. Yamaha's TR3 racing I engine actually has transfer ports that close 120-degrees before top center and I an intake port that opens and closes 94-degrees before and after top center, leaving only 26-degrees of crank-angle for pumping enough mixture to start the engine. I would think that this is something very near the absolute minimum even for an engine to be started by vigorous pushing of the motorcycle, and it would severely limit any efforts to improve this engine's power range through increases in transfer port-open duration.

Considering the intake-timing limitations imposed by the just discussed, it should be clear that the task of obtaining adequate crankcase filling in high-speed engines is not confined to establishing a suitable time-area value. Engines for motocross are restricted, in terms of port-open duration, by the need for a very broad-range output characteristic, to an intake period of not much more than 160-degrees. More than that virtually guarantees that they will be too peaky to be ride-able, or at least to be effective in terms of competitive laps times on most circuits, no matter what their time-area number may be. Road racing engines have an ultimate limit imposed by the starting problem already outlined. All of which means that you may regard the upper limit of intake-open duration for the former as being about 160- degrees, and about 200-degrees for the latter; about 80- and 100-degrees before and after top center. Exceed those limits, and the road racing engine will not start; the motocross engine's power band will narrow beyond the point of being useful no matter how impressive the maximum power figure may sound.

Intake-tract "tuning" will be vitally important no matter what kind of time-area is provided at the port window, and it is all too easy to get the pulsations out of phase with the piston by altering the intake timing. All alterations in intake timing should be followed with a careful check to determine if matching alteration of the intake tract length is not also required. Although this kind of work should be validated by actually running the engine with a stub exhaust attached, as outlined elsewhere in this book, a preliminary check may be run mathematically, using the formula for finding the resonant frequency of the necked flask formed by the crankcase and intake tract provided in the chapter on crankcase pumping.

For all of the reasons outlined in that chapter, which deals at length with intake-tract tuning, I have little confidence that the use of this formula will provide more than a rough guide as regards an engine's proper tuned intake length, but for some this rough guide may be all the guide they will have. It is better than nothing, if you compare the frequency thus derived for your engine's stock condition with the reality and make appropriate adjustments in the theoretically-obtained numbers for your modifications. It may also help to know that one of the better researchers in the field, Fujio Nagao, of Kyoto University, has verified that maximum air delivery occurs when the intake pipe's natural frequency provides a wave period 75-percent of the intake port open period. With all that, I still am inclined to believe that there is no substitute for actual testing, using a stub exhaust system to isolate intake effects.

ROTARY-VALVE TIMING

Blow-back during the period between top center and intake closing is the limiting factor in establishing time-area values for all two-stroke engines except those fitted with automatic intake valves (i.e. reed-valves). But the problem is much less severe and more easily managed when an engine has rotary-disc intake valving. Because of the considerable mechanical complication attending this method of crankcase filling, rotary-valves have not been used much in mass-produced engines, and the inconvenient width they add now seems unacceptable as quite good results can be obtained with either reed or piston-port induction. Be that as it may, the rotary valve still is best in terms of sheer engine performance, whether arranged for maximum power or for an ultra-broad power range. The rotary valve is free of the really serious blow-back problem afflicting piston-controlled valving, and it offers much less resistance to flow than reeds. People who are currently so infatuated with the reed-valve concept should consider that in the world of karting, where there is much more experience with both reeds and rotary-valves than motorcyclists have accumulated, the two types of engines have been separated into different classes. Why? Because while the reed-valve engines are inexpensive, they cannot match the performance of those with rotary-valves. So, even though the rotary-disc valve is for the moment out of fashion, the pressure of competition may eventually return it to the fore despite its handicap in cost and bulk.

Principally because it is so free of blow-back problems, the proper time-area of a rotary intake valve is much higher than for the piston-controlled port. This does not mean that one may indiscriminately whack away at the valve disc without getting into trouble, especially on the port-closing side of the disc cutaway. The valve's opening point has an influence on power output, but in general it is relatively insignificant as compared with port-closing. Curiously, the best port-closing timing for a very wide variety of disc-valve engines is about 65-degrees after top center, but before deciding to apply that timing you should consider that any appreciable delay in closing the port, after the piston has started down from top center, will cause some low - speed blowback. At higher engine speeds inertia effects in the intake tract will overcome the slight blowback caused by the delayed intake closing, but there will be a loss of low-speed power. All things taken into account, the best approach here is to increase the valve-closing delay in very small stages, not more than 2-degrees at a time, until the desired result is obtained. Remember that retarding the intake-closing point moves the engine's power peak higher, while reducing power at the lower end of the range. Remember, too, that changes in the length or diameter of the overall intake tract, such as would occur in substituting a carburetor of some different size, will alter the point at which port-closing delay reaches its optimum.

It is fortunate that rotary-valve engines are relatively insensitive to the point of intake-opening, because there is no clear pattern in existing examples to lend our efforts direction. If there is a rule, it is that the point of opening for rotary valves is best established right at the point of transfer-closing if you want a very broad range of power. On the other hand, maximum power is obtained by opening the intake port somewhat earlier: from 130- to 145- degrees before top center, which means that the intake and transfer ports have overlapping open periods. This presupposes that the engine will have been fitted with a proper expansion chamber. The diffuser section in such chambers is capable of returning a negative-pressure wave to the cylinder having a below-atmospheric value of something like minus-7.0 psi, and as this partial vacuum is communicated to the crankcase via the transfer ports, the pressure inside the crankcase itself will drop well below atmospheric, At lower-than-peak engine speeds, particularly, there is a tendency for this below-atmospheric crankcase pressure to equalize itself by drawing part of the charge back down through the transfer ports, which neatly cancels an equal and important part of the work done by the exhaust system. There is little that may be done to counter this in a piston-port engine, beyond working with an exhaust system proportioned to give a long-duration scavenging pulse that will maintain the negative pressure in the cylinder until after transfer-closing. But when the engine in question has a rotary valve, it is possible to open the intake side just as transfer passage back-flowing is about to occur, and balance the pressures with mixture drawn in through the carburetor instead of robbing from the cylinder. As you might guess, this neat trick doesn't work properly unless the intake-open period is delayed until after pressure in the crankcase has been pulled down to atmospheric or below - which means that intake-open timing is very closely tied to the exhaust system's pulsing and the overall flow characteristics of the transfer ports. A few hours work with an oscilloscope and pressure-transducers would get you right on target, because you would then know with a high degree of certainty the precise point at which crankcase pressure did in fact fall to atmospheric, but few people have that very expensive equipment and most will have to achieve the same result through a laborious process of cut-and-try. The same kind of cut-and-try in fact, as is needed to locate the optimum point for intake-closing.

All of the major influences on time-area requirements have been covered. There are others that could be major, but for practical reasons are not. Crank case compression ratio is one. Both intake and transfer time-area requirements do vary in inverse proportion with crankcase compression ratios, simply because as crankcase pressures are reduced so are the pressure differentials that cause gases to move. You are spared having to worry overmuch about this factor because virtually all modern engines have primary (crankcase) compression ratios very near 1.5: 1 - this having proven to be best for almost every application, and is mechanically easy (it is what you get with flywheels and crankcase of normal proportions). Exaggerated angles of entry into the cylinder could also create ports with window areas misleadingly larger than the passages themselves, but again this condition- while always present in some degree - is seldom serious enough to warrant one's worry while working time-area calculations. Finally, extraordinarily poor casting techniques and/or badly misshapen port cavities could lend ports such low coefficients of flow as to upset one's calculations, but one almost never sees outstandingly bad work from any of the major manufacturers. All of their casting work is quite good, and will not be much improved by even the most painstaking carving and polishing on your part -which may be discouraging, but does tend to preserve the validity of the time-area values I have presented here.

CRANKCASE PUMPING

Back before the unpleasantness of the early 1940s, a number of two-stroke motorcycle engines were built with external scavenging-air pumps – much in the fashion of today's GMC truck engines -despite the resulting penalty in bulk and manufacturing cost. It was then thought that no engine relying on ordinary crankcase pumping for scavenging-air delivery could hope to compete with the better four-strokes in terms of specific power output, and there was every reason for that pessimism: For one thing, no one in his right mind would design a piston-type air pump with as much clearance volume as inevitably exists in an engine's crankcase. Neither could it be considered desirable to pre-heat the scavenging air before it is delivered to the cylinder, which is precisely what happens in the crankcase-scavenged two-stroke engine. Finally, using the two-stroke engine's crankcase as a scavenging pump condemns the connecting rod bearings to a diet of too much raw fuel and too little oil. Only the fact that resorting to a separate, external scavenging pump roughly doubles the cost of a single- or twin-cylinder engine, making it more expensive than a four-stroke engine of equivalent power output, has kept us at work on the crankcase-scavenged two-stroke. And over the years, out of necessity, we have learned to make the crankcase function as a pump with an effectiveness that would astonish the engineers of fifty years ago.

Much of the improvement in the air-delivery capabilities of crankcase pumping can be traced back a half-century to a two-stroke stationary engine (used to drive an electrical generator) that neither employed an external scavenging pump nor relied upon its crankcase to do the job. Instead, the work of moving air through the cylinder was performed by the effects of sonic waves and inertia in the engine's intake and exhaust pipes. The former was connected directly to the cylinder's scavenging ports; the latter was a conventional, if lengthy, straight pipe. This engine had to be motored up to its operating speed and a blast of compressed air directed into its intake pipe to start it firing. But then a combination of wave and inertia activity in its intake and exhaust pipes would take over, to scavenge and recharge the cylinder, and the engine would thud-thud merrily along at that speed until it ran out of fuel or broke. Being utterly dependent on the resonant frequencies of the attached plumbing, it would of course run at only one speed-but that is a virtue, rather than a disadvantage, in an engine used to turn a generator. And while the engine described was neither very powerful for its size nor particularly efficient, it was simple and trouble-free... and pointed the way for the hyper-powerful racing engines of another, later time.

Obviously, engines capable of running at only one speed, and which must be cranked up to that speed before they will start, are of limited usefulness in the context of motorcycling (although a couple of road racing engines approaching that condition come to mind). Even so, few of the current crop of high-performance engines would perform at all well without help from resonant pipes. Rotary-valve engines function very nicely without much assistance from intake-tract resonance, but those using piston-controlled intake ports -which necessarily have their intake-open period spread symmetrically before and after the piston reaches top center - tend to regurgitate a lot of the mixture drawn into the crankcase by the ascending piston unless this backflow is in some manner prevented. Combined inertia and wave activity, which may attain pressure values in excess of 1.5 atmospheres, are enough to balance any pressures created by the descending piston - even if port closing is delayed until as much as 100-degrees after top center, as is the case in highly-developed road racing engines. Of course, these effects are extremely transitory, and will prevent backflow only if their arrival is properly timed.

RESONANCE EFFECTS

Timing wave arrivals on the two-stroke engine's exhaust side is a relatively simple task, as the exhaust system's resonate frequency is almost exclusively a function of its length. A far more complex situation exists on the engine's intake side, for there you have not straightforward "organ pipe" resonance, but a resonating flask consisting of the crankcase and intake tract. As it happens, there is a fairly simple formula for determining the resonant frequency of flasks, which is

$$F = \frac{V_s}{2\pi}\sqrt{\frac{A}{V_c\left(L + {}^1/_2\sqrt{\pi \times A}\right)}}$$

Where V_s is sonic speed (usually about 1100 ft/sec)
A is the cross-sectional area of the inlet
L is the inlet pipe length
V_c is the flask (crankcase) volume

From the formula, you will see that resonance in a flask, or in an engine's intake system, (which includes the crankcase) , frequency is directly proportional to the square root of intake pipe cross-sectional area; inversely proportional to the square root of crankcase volume; and also inversely proportional to the square root of intake pipe length. Researcher Fujio Nagao, of Kyoto University, who has done much of the basic investigation in this field, has established that intake tract length for maximum air delivery should be that which provides 75 pressure fluctuations "coincident with the period of inlet port opening". That is so say, if an engine's intake period in degrees is 120 degrees and its torque peak is at 6000 rpm, then its intake period in time will be 0.003-second and the intake system's frequency coincident with 75-percent of that, or 125 cycles per second.

Unfortunately, this straightforward picture begins to get very complicated as one tries to apply any of the preceding to the concrete example. We can, for instance, only guess at sonic speed, as it is dependent on temperature and one can only guess at the temperature of a stream of gases simultaneously being cooled by evaporating gasoline and heated by surrounding metal. But that is not the greatest difficulty one faces in calculating the proper length: The flask in question here has a volume that varies continuously with the piston's motions, and the intake tract is in most engines far from being a simple tube (not only area but shape will vary from the carburetor intake bell to the port window). Moreover, the inertia of the fast-moving column of gases in the intake tract must also have an effect. In short, calculating intake tract length is a highly complex problem, and while a trained engineer with a heavy background in higher mathematics could probably do the job given sufficient time, it really is a task for the computer - and even then it will be necessary to use abbreviated formulae to avoid accumulating too many expensive computer hours on the problem.

In all probability, those who will read this book will have neither a computer nor the spare hours for a theoretical determination of intake tract length. Thus, it becomes necessary to arrive at this length experimentally, which fortunately is a much shorter (and more accurate) means of accomplishing the same thing. The first step in that direction is to isolate the influence of intake tract length on the engine, which means removing any effect the exhaust system may have on the results obtained. To do this, you will have to fit your engine with a stub pipe too short to be effective within the engine's projected operating range, yet long enough to prevent the "back-breathing" of air in through the exhaust port to upset mixture strength and thus mask intake-length influence. For small-displacement cylinders, a total exhaust length of 3 1/2-inches (measured from port window to the end of the exhaust stub) should be satisfactory. Middle-displacement engines should have a 4-inch stub-exhaust, and those with a 350-400cc cylinder displacement a length of 4 1/2-inches. Quite obviously, the sheer noise level produced by stub-exhausts will make some form of muffling a necessity in many areas, and where such is the case you can add muffling without upsetting the experiment by introducing a large-volume chamber into which the exhaust-stub empties. This chamber should be as large as can be fitted on the motorcycle without making it impossible to ride, and the attached muffler should have sufficient internal diameter to prevent any backpressure from developing in the system. The hardware that makes up this no-effect exhaust system need not be beautiful, nor tucked in on the motorcycle neatly enough to permit anything more than straight-line riding, but it should be easy to install because you will have to re-tune the intake length with every change in port timing and/or carburetor diameter. There will, of course, be a change in the system's frequency with every change in crankcase volume as well, but such changes will be too small to worry about in most instances.

For intake-length experiments, you will need not only the stub-exhaust system, but a tachometer on your motorcycle and either a long, straight road (or trail) or a hill. These last are required because you must load the engine heavily enough to allow you a good, seat-of-the-pants reading of the engine speed where the pipe has its effect on power. And that effect will be unmistakable, for the engine will pull very strongly when it comes "on the pipe". You will also find that intake pipe length can move the stub exhaust-equipped engine's power peak over a very wide speed range. My friends at McCulloch, who acquainted me with the stub-exhaust testing technique, tell of tests they have performed with an engine having an intake period of 120-degrees, and with pipes ranging from 5- to 9 $^1/_2$-inches in length they were able to move the power peak anywhere from 3000 rpm to 8000 rpm. The 9 $^1/_2$-inch length gave substantially the same power at 3000 rpm and 4000 rpm; at 5000 rpm, an 8 $^1/_2$-inch was best; at 6000 rpm and 7000 rpm the same power could be had with either a 9 $^1/_2$- or 6 $^1/_2$-inch length; and the same was true at 8000 rpm, but with lengths of 5 $^1/_2$ and 8-inches. From these tests, it was determined that the best compromise length for the entire range 3000-8000 rpm was 8-inches; a 5 $^1/_2$inch length was slightly better for the range 6000-8000 rpm. It should be understood that these lengths only apply to this specific example, and would vary with any changes in intake timing, crankcase volume and intake tract diameter. They are given here only to illustrate that: A) it is possible to make a given intake timing work over a wide speed range by varying intake tract length; and B) that the same length may produce more than one peak, or conversely that more than one length may be effective at any given power peak.

CARBURETOR LOCATION

When performing your own experiments with intake tract length, you will be tempted to mount the carburetor as close to the cylinder as possible, and make adjustments by varying the length of an intake stack added on the carburetor's mouth. Don't do it! That certainly is the most convenient method, but a carburetor placed too far back toward the intake port window is going to be subjected to radical pressure fluctuations, due to wave activity in the system, and those pressure fluctuations do terrible things to the carburetor's ability to meter fuel. Worse, the largest departure from the desired mixture strength will occur precisely at the engine speeds where intake ramming is strongest, which makes the task of selecting a main jet impossible. So the carburetor must be located out at the end of the intake tract, rather than close to the cylinder- which is unfortunate, as maximum air delivery into the crankcase is obtained when the reverse is true. Perhaps some form of fuel injection is the answer.

CRANKCASE PUMPING

Mixture-strength problems can also occur due to the plumbing between the carburetor and air cleaner, and it should go without saying that such plumbing may also add to the intake tract's effective tuned length. The common practice of connecting carburetor mouth and air cleaner with a section of rubber hose may have much to recommend it from the standpoint of convenience; it also is likely to establish either an extension that becomes part of the effective tract length, or a secondary resonating system that heterodynes at some frequencies with the main tract and thus upsets its proper functioning. Therefore, it is good practice, if sometimes inconvenient, to make connecting plumbing both as short and as large in diameter as is possible. Air cleaners having plastic-foam elements may be mounted very close to the carburetor without penalty, but the paper-type filter may, if it is too close to the carburetor mouth, become saturated with oil and fuel - in which case it will refuse to pass any air at all.

CRANKCASE VOLUME

At one time, it was every two-stroke tuner's habit to begin any serious attempt at extracting more-than-standard horsepower from a given engine by increasing the engine's primary compression ratio - that is to say, the ratio between crankcase volume with the piston at top center, and at bottom center, as in the following expression:

$$\text{Primary compression ratio} = \frac{\text{case volume at TDC}}{\text{case volume at BDC}}$$

$$\text{Or, } CR_p = \frac{V_1 + V_2}{V_1}$$

Where, CR_p is primary compression ratio
V_1 is crankcase volume at BDC
V_2 is piston displacement

I suspect that the popularity of this practice, "stuffing" crankcases, was derived from the fact that the old T-crank Villiers was then the two-stroke engine most frequently being modified. This engine was built more with an eye toward manufacturing cost and long-term reliability than power output (of which it had only extremely modest amounts) and the configuration of its crankshaft and crankcase provided only a very low primary compression ratio. Too low, in fact, for anything even remotely approaching high speed operation, so that any efforts at increasing its primary compression ratio were immediately reflected in a power increase. But a lot of people simply assumed that if increasing the Villiers engine's primary compression ratio from, say, 1.2: to 1.4:1 was good, then raising it even further would be better -and that all two-stroke engines would benefit from being given the same treatment. There, they fell into error.

Referring again to the work of Fujio Nagao (and his results have been verified by other researchers) we find some very interesting conclusions with regard to primary compression ratios: First, Nagao tells us that for given port areas, there is an engine speed at which maximum air delivery to the cylinder occurs, and that this engine speed is inversely proportional to crankcase volume, but that the maximum value changes only slightly with changes in crankcase volume. To put it another way, the crankcase-pump's volumetric efficiency is nearly constant, but the engine speed at which it attains maximum efficiency rises as crankcase clearance volume is reduced. Significantly, too, Nagao goes on to say that any deficiency in air delivery due to a crankcase volume too great for a given engine speed is fairly well compensated by properly tuned intake and exhaust pipes. He says in conclusion that "little advantage is obtained by making the crankcase volume excessively small."

Later work in the same field indicates that Nagao understated the case, and that there may well be positive disadvantages in excessive reductions of crankcase volume. Hiroshi Naito, who I believe was responsible for the fantastic Yamaha GP racing engines, has indicated that there is little improvement to be had even from ultra-high speed engines with primary (crankcase) compression ratios above 1.5:l. This apparent anomaly can be explained by studying the effects of the whole scavenging system, from the intake tract through the crankcase and scavenging ports and right on out into the exhaust system. And we must think in dynamic, rather than static terms. Starting with the exhaust system, we find that it is possible to evacuate the cylinder to well below atmospheric pressure by using the "extractor" effects of the expansion chamber's diffuser. If the lengths and proportions of the exhaust system are properly established, the fresh charge moving up through the transfer ports will not only be pushed through by the pressure below but will get additional aid from what is, in effect, an exhaust-produced vacuum in the cylinder. Further, this vacuum may well be communicated down into the crankcase, via the transfer passages, and crankcase pressure "trapped" at a below atmospheric value when the transfer ports close. This factor can be very important, as it produces an air-delivery volume greater than would be possible with crankcase pressure alone to impel the fresh charge into the cylinder. But that isn't the whole story, obviously, for the "pull" on the exhaust side of the cylinder is matched by a "push" where mixture from the carburetor enters the crankcase - with a force equal to about 1.5 atmospheres in advanced engines. Thus, we have evacuated the crankcase to something below atmospheric pressure before refilling it with a fresh charge, and the crankcase will have been charged to something above atmospheric pressure by the pulsations in the intake tract. All this has implications in terms of crankcase volume, for if we assume that the positive and negative pressures applied remain constant, then bulk flow through the cylinder will increase with increases in crankcase volume. Does that seem obscure? Then consider that there is more air in a 1000cc flask compressed to 1.5-atmospheres than in one of 500cc capacity. In light of this, you might assume that it is time to start reducing crankcase compression ratios - especially as pumping losses (horsepower absorbed in doing this work) rise as to the third power of compression ratio. Actually, this isn't quite true either, for reductions in crankcase compression ratio cause an equal reduction in the amplitude of wave activity in the intake tract, which in turn reduces the ramming pressures available to charge the crankcase. So, crankcase compression ratios must be established to balance the conflicting requirements of volume and wave strength, and if it is pure, peak horsepower that interests you, then you will find that a primary compression ratio of 1.5:l represents something very near the optimum. Of course, this will only be true if carburetor size also is selected with an eye toward maximum horsepower.

Broad range performance is another matter entirely. For motocross, you will require much more flexibility from an engine than would be true of road racing or even flat-track. In the large-displacement classes particularly, where sheer engine size makes it altogether too easy to get more horsepower than a motorcycle's rear tire can apply to the ground, an engine's ability to pull strongly over a very wide speed range is much more important than any peak reading. For such applications, the best engine is likely to be one with a primary compression ratio well below 1.5:1, a long, slow-taper expansion chamber and a relatively long intake tract. With regard to this last item, I might add that it is possible with intake lengths of about 30-inches to boost an engine's crankcase delivery ratio (volume of air pumped, per cycle/piston displacement) to as much as 1.2:l. However, due to restrictions in available time, these ultra-long pipes will not work except at very low engine speeds and, moreover, tend to work only over a very narrow :peed range. Indeed, all tuned intake pipes effectively reduce an engine's operating range, though this is compensated by their providing a power boost at some engine speeds, and it might be that a near-zero length would provide the best spread of power. The difficulty here is that some sort of smooth passage must be provided for the transition from the substantially-square intake port window to the round carburetor throat, and if you add the length of that passage to the carburetor's length, then you have a resonating intake system even if you don't want one. The choice is thus reduced to selecting a length that provides the best results overall.

There is one means of halting resonant effects in the intake tract, and to accomplish that you have only to reduce the carburetor throat in diameter until its cross-sectional area represents 35-percent, or less, of the intake port area. Curiously, a carburetor of that small size placed at the port window, behind the intake pipe, will not stall wave activity in the system- but one located out at the end of the intake pipe acts in effect as a wave-damper. I mention this only for its value as a curiosity, as there is no point in fitting such a small carburetor - except, perhaps, on a trials motorcycle - and if fitted at the port window it would be subject to the same mixture-strength instability as any other carburetor preceded by a length of resonating pipe.

REED VALVES

Another, indirect means of gaining the benefits of a resonating, ram intake tract (as well as certain others) is to interpose a check-valve between carburetor and cylinder - which will insure a one-way flow, and trap in the crankcase anything that passes through the valve. Now as it happens, the only check-valve currently capable of functioning fast enough to keep pace with a high-revving two-stroke engine is the one generally called a "reed-valve". The name fairly accurately describes the valve, which has a thin, broad metal or phenolic reed seating against an aperture and clamped at one end so that it may bend away from said aperture. There usually are multiple reeds and apertures in any reed-valve assembly; also in most instances these will be set at an angle oblique to the direction of gas flow, so as to minimize flow losses past the reeds. Usually, too, there will be four separate reeds, clamped against the sides of either a pyramid or a wedge, with the carburetor feeding the interior and the point being aimed into the crankcase. At one time many attempts were made to use comparatively stiff reeds, to overcome a tendency for the reeds to fracture and drop off at high engine speeds, but now everyone utilizes thin reeds backed by stops. The stops, which are simply thick, curved strips of metal clamped over the reeds, control both the reeds' travel and the shape of reed-

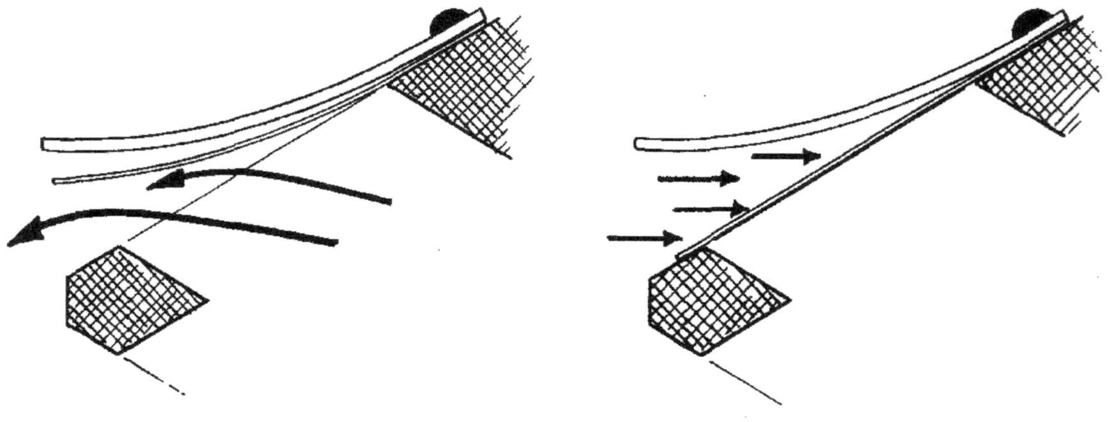

A reed valve's reeds usually are used in groups of four or more and positioned at an angle to the direction of flow to improve efficiency, with curved stops to prevent over-stressing the reeds.

bending. You will appreciate that a reed held only at its end will bend quite abruptly right next to the clamp that holds it in place, and that it may well bend too far for its own well-being under some conditions. Reed-stops prevent both of those things from happening, forcing the reed to curve very evenly around the stop and allowing it to bend only as far as the fatigue-resisting properties of the reed material enable it to withstand.

Only rarely, today, do reeds fracture at their root and drop into an engine's crankcase - but it still happens, and for that reason many manufacturers who build reed-valve engines prefer to employ plastic (usually a fiber-reinforced phenolic sandwich) reeds. Steel reeds can do terrible damage to an engine's interior when they break-off and are aspirated up through its transfer ports; the phenolic reed simply gets gnawed into fragments and expelled out the exhaust port. This danger, with steel reeds, apparently is very real, for there can be little doubt but that engine performance is improved by the use of steel over phenolic in the reed material - simply because steel is the more flexible material.

Actually, the failure of a complete steel reed is an extremely rare event. Much more frequently the failure will be confined to a piece breaking away at the reed's tip - which is caused by the repeated impacts of the reed against the reed block. For many years this problem seemed to defy solution, as very thin reeds were not strong enough to resist these fretting fracture, and thicker reeds were shattered by their own greater inertia. Yamaha found the solution, which is beautiful in its simplicity. The 1972 Yamaha reed-valve (they call it "Torque Induction") engines have reed blocks coated with neoprene, and while the coating is very thin, it has enough resilience to cushion the impact of the reeds and prevent fretting failures. I understand that the Yamaha reeds are of stainless steel, with thicknesses of 0.008- and 0.006-inch for the 250cc/360cc and 100cc/125cc/175cc engines, respectively. The use of a stainless steel as a reed material will be explained if you consider the devastating effect of even a very little rust on such thin strips of metal.

I am much inclined to doubt that reed-valves will become popular in road racing engines, as they do introduce a flow restriction into the intake system that must have a depressing effect on peak horsepower at ultra-high engine speeds. But there is every reason to assume that "Torque Induction" will come into very wide use for every kind of off-road motorcycle: There is, for example, the reed-valve's ability to make the most out of positive intake-resonance effects while stalling out-of-phase resonance. Also, the reed-valve has, by its very nature, the ability to effectively adjust intake timing to suit all engine speeds. Finally, it does seem that reed-valving eases problems with broad-range

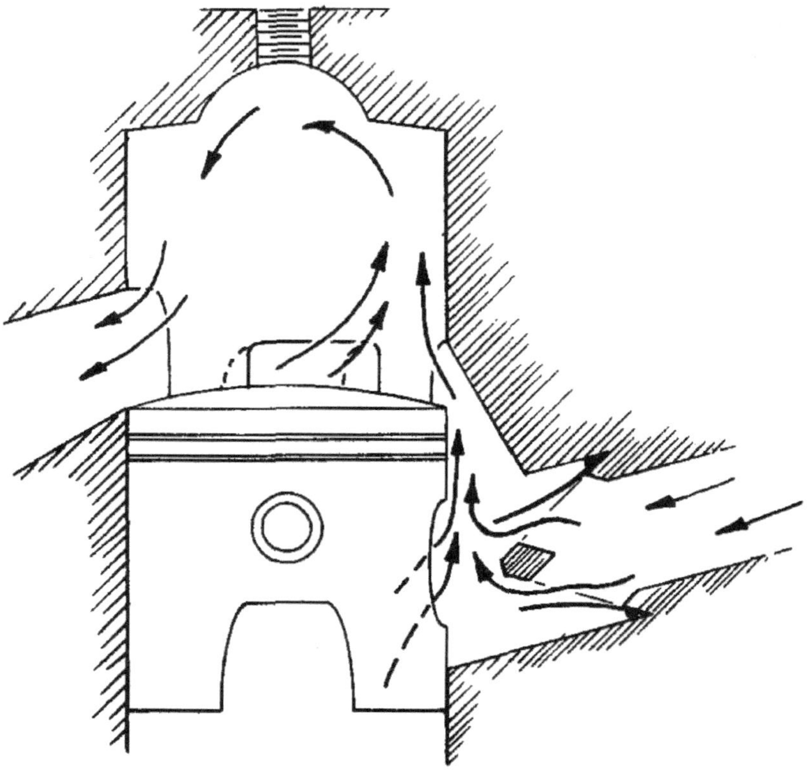

Reed valves always restrict intake flow, but this may be compensated in part by taking advantage of their automatic inlet characteristics and carving a boost port in the rear cylinder wall.

carburetion, ending the oft-noted tendency for carburetors to produce wildly-varied mixture strengths at different engine speeds.

Another benefit that accrues with reed-valving is that with all possibility of back-flowing out the intake removed, it becomes possible to use the rear cylinder wall for something other than a support for the piston. Yamaha, long an advocate of multiple transfer ports, has added another port, opposite the exhaust port, in its reed-valve engines. This port is as high as the other (four) transfer ports, but is very narrow. Significantly, this extra transfer port is not fed mixture from the crankcase; it relies entirely upon the extractor effect of the exhaust system to pull mixture from the carburetor through the reeds and reed cavity, and up through the port into the cylinder. My friends at McCulloch - who have an enormous accumulated experience with reed-valve engines, tell me that power characteristics can be altered very radically by widening and narrowing this boost-port, and/or by arranging for mixture compressed in the crankcase to flow up through the boost port instead of relying purely on the depression created in the cylinder by the exhaust system to do the job. But they also seem to think that the boost-port's width is something one determines almost entirely through experimentation, so it would appear to be wise to start with a very narrow port window and widen it gradually until the desired power characteristics are obtained.

THE ROTARY VALVE

Better than reed valves in function, if not in mechanical convenience, is the rotary valve. Specifically, the disc-type rotary valve, which is the only variety currently in use in motorcycle engines. There is little point in dwelling on this kind of intake valving over-much, because it really is only practical on a GP road racing engine. Touring-type engines, if they have more than a single cylinder (and the trend obviously is toward multis) become inconveniently wide when a disc-valve assembly is added at each end of the crankshaft, and this type of intake valving is all but impossible to apply to an in-line engine with more than two cylinders. The rotary-valve's extra width is, of course, an embarrassment even on single-cylinder engine in off-road applications.

If you ignore the mechanical disadvantages of the rotary-valve, it becomes highly attractive from the standpoint of not only peak power, but broad-range power. The reason for this is that it does not leave the engine to rely upon pressure waves marching back and forth in a pipe to prevent charge loss back through the carburetor. It is possible to extract just as much horsepower, at the peak, from a given engine with either rotary-

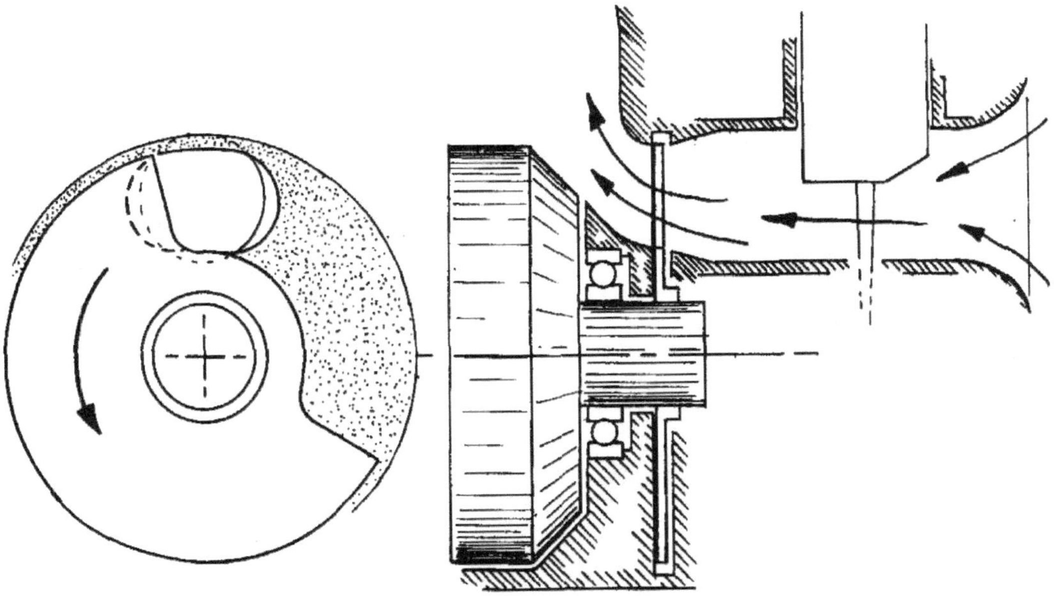

The disc-type rotary valve can impose penalties in cost and bulk, but provides the best known solution to the intake problem, yielding both maximum power and a very broad power range.

valve or piston-port induction, but the latter will lose power very rapidly either side of the bhp-peak while the rotary valve continues to function and crankcase pumping remains effective.

INTAKE PORT SHAPE

Intake port timing is covered elsewhere in this book, but this certainly is the place for a discussion of intake port shape - which has a definite influence on crankcase pumping efficiency. In general, the best flow coefficient for any given timing-area value will be obtained with the widest-possible port. That is to say, a port that wraps around the cylinder as far as is permitted by mechanical considerations (stud placement, transfer port location, etc.). Of course, with a very wide port there is a tendency for the rear edge of the piston skirt to snag at the bottom of the port window, which means that it may be necessary to use a window shape more nearly round than square to prevent rapid wear at the bottom of the piston skirt. Which is often good practice in any case. A rounded port window, or one with a V-shape to its lower edge, provides what effectively is a slower rate of port opening, which is very useful in reducing intake roar - a point that must be

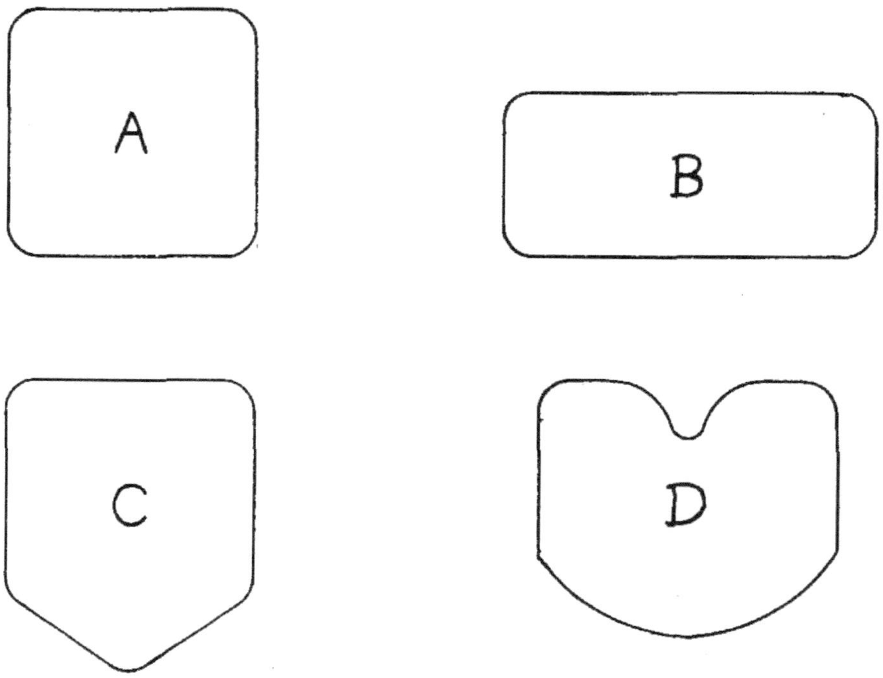

All of the port window shown here have the same area, but "B" has a better flow coefficient than "A"; intake roar is reduced by the notched floor of "C"; and "D" has a partial ring-restraining rib above and a floor that prevents piston-skirt snagging.

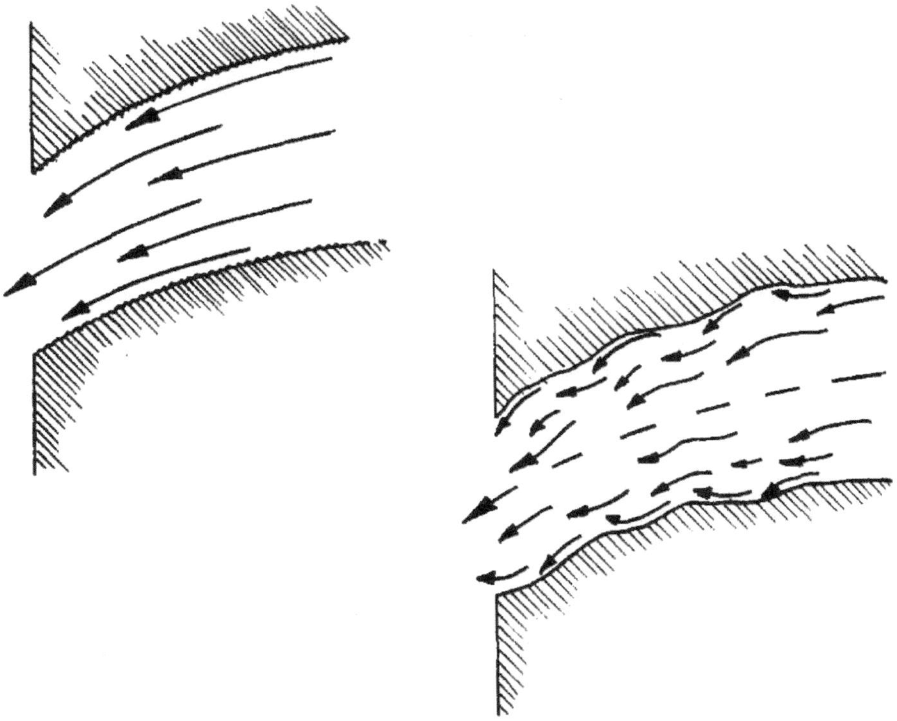

Air flow through ports is largely undisturbed by surface roughness. Uneven walls, with ripples, disrupt flow even when very carefully polished to mirror-like brightness.

considered in a time when statutory noise limits are popping up all over the place. Also, the more-gradual opening of the port tends to extend the duration of the sonic wave that is used, on its return trip, to "supercharge" the crankcase, and that has the effect of broadening an engine's power band. Finally, a port with rounded corners has a much better flow-coefficient than one that is square. The same may not be said for rounding back the lower edge of the piston skirt, as that extends the intake timing- to permit backflow as the piston descends -without producing any measurable improvement in flow coefficient. It is, however, possible to improve flow with a down-turned lip at the top edge of the intake port window. But the primary thing one must remember when carving away at an intake port is that ripples in the port walls, or any sudden change in cross section, have a far more damaging effect on flow-coefficient than a slightly rough finish in the port. Therefore, it is vastly more important to smooth the port than to give it a mirror-finish. And it should be obvious that the port face, the gaskets and heat-block (if any) and carburetor should all align very neatly, without any steps between parts - even if that means doweling everything in place to assure alignment.

CYLINDER SCAVENGING

Scavenging, in the context of piston-type internal combustion engines, is the process in which the products of combustion are cleared from a cylinder at the end of the power stroke and a fresh air/fuel charge is introduced in preparation for the compression and power strokes to follow. This process is common to all Otto-cycle engines, but it can be accomplished in two entirely different ways: In the four-stroke cycle engine, it occupies at minimum a full 360-degrees of crankshaft rotation, with one piston stroke being devoted to pushing exhaust products from the cylinder, past a valve in the cylinder head; the return stroke aspirates a fresh charge in through another port and past another valve. Thus, there is a fairly complete mechanical separation of the gases involved, and while valve timing will commonly be arranged so that there is some overlapping of the exhaust and intake phases of this operation, little dilution of the fresh charge by exhaust gases is possible, and any short-circuiting of this air / fuel mixture out the exhaust port during the early stages of the intake-open period has no adverse effect on the weight of the charge ultimately trapped in the cylinder at intake-closing. Obviously, the mechanism required to operate a four-stroke engine's valves adds considerable complication to the basic crank-rod-piston assembly, but the very efficient scavenging obtained is, for most applications, considered to be worth the valve-gear complexities. And the very high brake mean effective pressures resulting from this high scavenging efficiency - a bmep in the order of 200 psi for highly-tuned examples of the type - offset the scheme's single disadvantage, which is that power strokes occur at 720-degree intervals.

Two-stroke cycle engines deliver power strokes twice as often, at 360-degree intervals, and in those intended for industrial (trucks, electrical generators, etc.) or marine applications essentially the same bmep as are obtained from four-stroke engines are to be expected. But in such engines one finds an even greater mechanical complexity than in the four-stroke design, for in all two-stroke engines the scavenging process occurs in time borrowed from the compression and power strokes. In effect, this means that all of the cylinder clearing and recharging for which 360-degrees of crank rotation are reserved in the four-stroke engine must occur while the piston is halted at the bottom of its stroke. And, lacking time for a more leisurely exchange of gases, the process must be helped along by extremely large port areas and high scavenging pressures. Usually, engines of the type being discussed will have multiple exhaust valves in their cylinder heads, and a ring of windows around the cylinder's base through which scavenging air is forced by an engine-driven pump. Such engines almost invariably are diesels, in which fuel is injected only after all the valves and ports are closed (injection beginning at TDC and continuing for perhaps 70-degrees of crank angle) and the scavenging pump delivers air in excess of what is required to fill the cylinder, so there is no loss of charge nor any dilution to cause a loss of power. Unfortunately, not only is this type engine very complicated and expensive, it cannot be scaled-down to a size useful in motorcycle terms - simply because the exhaust valves will not open and close fast enough to keep up with the kind of crankshaft speeds needed for the power outputs we have come to expect from our small-displacement engines.

In the end, only the familiar "piston-port" transfer and exhaust valving arrangement is suitable for motorcycle engines, and that is -with an exception, of sorts, existing in the use of disc- or reed-type intake valving –precisely what has come to be universally applied. Present motorcycle engines are all scavenged through windows in their cylinder walls, with scavenging air being supplied from their crankcases. This system is beautiful in its simplicity, but it does have serious short-comings: First, there is the relative incapacity of the crankcase as a scavenging-air pump, which prevents even the hope of having excess air to use in clearing the cylinder. Secondly, the use of the piston's motions to open and close (actually, to uncover and cover) the exhaust and transfer ports creates enormous difficulties in a number of areas related to clearing and recharging the cylinder. The low scavenging pressure available makes it absolutely essential that pressure in the cylinder be no more than slightly higher than atmospheric when the transfer ports open, which means that the exhaust phase must begin well in advance of the uncovering of transfer ports. And, because the piston-controlled exhaust timing is necessarily symmetrical, the exhaust port will remain open long after the transfer ports close - leaving an unobstructed opportunity for the fresh charge to escape the cylinder. Indeed, the charge injected into the cylinder has every reason to escape, as the upward motion of the piston, moving to close the exhaust port and begin the effective compression stroke, is displacing the gases above its crown. Gas pressures always try to equalize, and those in the cylinder can only do that by moving back into the transfer ports, while these are still open, and out the exhaust ports. Thus, it is virtually inevitable that some portion of the fresh charge will be lost into the exhaust system, and that the upward stroke of the piston will also tend to aspirate some of the charge back down in the crankcase.

Difficulties inherent in the piston-port scavenging system are not confined to charge loss, or backflow into the crankcase. One of the great problems is created by the lack of mechanical separation of the exhaust gases and the incoming fresh charge. We expect that the engine's exhaust gases will choose to escape from the exhaust port, and that the charge coming in through the transfer ports will push the residual exhaust products ahead of it to completely clear the cylinder, but the actual process is by no means that tidy. The cylinder pressure may drop very neatly to atmospheric, or even below, but it still will be filled with exhaust gases, and these will not necessarily be swept out the exhaust port merely because other gases have entered the cylinder. In point of fact, it is possible to short-circuit the scavenging flow straight from the transfer ports to the exhaust port and leave the exhaust residuals in the upper cylinder entirely undisturbed. This possibility has haunted the design engineer throughout the two-stroke engine's long history, and many an elaborate system of ports and piston-crown convolutions has been created to confound this worst of all demons.

For many years cross-flow scavenging was preeminent, principally because it makes maximum use of cylinder wall area. In the cross-flow engine, ports ring virtually the entire lower cylinder, with half being exhaust and those opposite being transfer. Given the direction of the flow emerging from the transfer ports, the charge would shoot straight across and out the exhaust side but for the shape of the piston crown - which lips up on the transfer side to form a deflector that redirects the transfer flow upward. Cross-flow scavenging is still employed in outboard-marine and model airplane engines, in the latter because it is a manufacturing convenience and in the former because its insensitivity to scavenging pressures and volume of flow provide superior low-speed, part-throttle running characteristics. The cross-flow system is, however, handicapped in terms of maximum power by the large surface area created in all that lumpiness in the

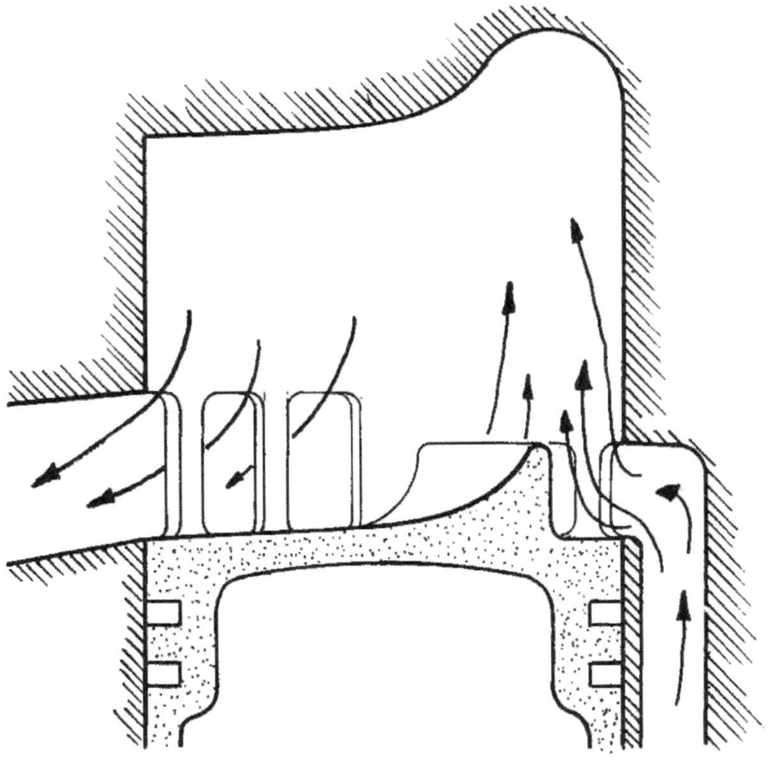

Cross-flow engines have large port areas and bulk scavenging flow but too much piston crown area to permit high specific output.

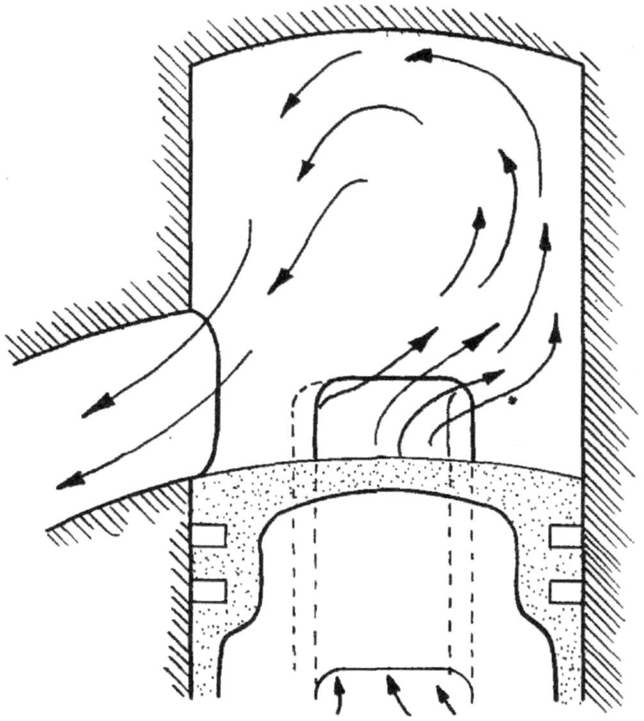

Dr Schneurle's "loop" scavenging system and its varients have been almost universally adopted by makers of high output two-strokes.

piston's crown, which very greatly increases the heat flow into the piston and lowers the compression ratio a given cylinder will tolerate as compared to a piston with a flat or slightly domed piston crown. Apart from this thermal problem, there is much to recommend cross-flow scavenging, but the thermal problem is of sufficient magnitude- even in water-cooled engines - to remove it from serious consideration for any high-output, two-stroke engine.

Numerous scavenging systems not requiring a deflector-type piston have been tried: The pre-WW2 Villiers had exhaust ports on opposite sides of its cylinder and four transfer ports, in pairs, between them. A Barnes and Reinecke design had a ring of exhaust ports located above a ring of transfer ports and a part-conical piston crown, all of which sent the scavenging flow in a narrow column up the middle of the cylinder, and forced the exhaust outflow to follow a path down the cylinder walls. Curtiss employed multiple transfer and exhaust ports on opposite sides of a cylinder, and biased the direction of the transfers upward and to one side, so that the fresh charge spiraled up into the cylinder. But the best of the scavenging systems was one devised by a Dr. Schneurle, of Germany, in which a pair of mirror-image transfer ports flanked a single exhaust port, directing the scavenging flow toward the cylinder wall opposite the exhaust, and upward, to loop over and thus clear the cylinder. Schneurle's loop-scavenging method was patented by him, in 1925, and this had the effect of simultaneously elevating German industry's fortunes in the two-stroke engine field while forcing practically everyone else to seek alternative and less-efficient systems. Of course, now that the Schneurle patents have expired everyone employs some form of his scavenging method, although best results are being obtained with more than Schneurle's original pair of transfer ports.

Only a decade past, East Germany's MZ was considered to be the repository of really advanced research in high-speed two-stroke engine design, and one Walter Kaaden could be said to have the best grasp of the intricacies of scavenging systems of anyone working in the field. Today, no discussion of two-stroke engine scavenging is possible without concentrating almost exclusively on development in Japan. Japanese engineers did not invent the two-stroke engine, nor have they employed any system of scavenging ports that has not seen earlier service elsewhere. But they have done an enormous amount of basic research directed at quantifying what previously has been known only in terms of generalities; they have established very firm design criteria for the management of factors that once were decided almost purely through cut-and-try experimentation. Of course, none of this would be of more than incidental interest but for the fact that some of the Japanese firms have abandoned their once-absolute policy of secrecy and are sharing what they have learned with the rest of the world. Yamaha, particularly, has made a vast contribution to the overall state of the art by publishing fairly specific criteria for the port timings and areas required for engines of any given cylinder volume and operating speed. Like many others, I knew that port timing and area were interrelated factors, but the job of obtaining and sorting through data on a wide range of engines to establish a pattern, and then experimentally verifying conclusions was too time-consuming and expensive to even contemplate, as an individual. Yamaha has done that work for us, and published enough information on the subject to complete at least my understanding (a detailed discussion is presented elsewhere in this book as a chapter, headed, "Port Timing"). From a number of SAE papers from Japan - as well as examples from Germany and the United States -and my own experience, I have also accumulated much incidental information related to the shapes, number and disposition of ports. These factors profoundly influence scavenging flow, which influences horsepower very greatly, and we will for the moment concentrate on them alone.

EXHAUST PORT

The difference between success and failure with a modified engine can be the treatment of the exhaust port. Even assuming that no change is made in exhaust-port timing, simply widening the port window will result in a power increase; it also can result in drastically-shortened ring life, amounting in extreme examples to outright and nearly-instant breakage of the rings, and/or severe overheating of the piston crown. There are reasons for these problems: A two-stroke engine's piston rings always bulge out into any port window they pass, and while transfer port windows seldom are wide enough to permit this to an extent sufficient to cause difficulties, the same certainly may not be said of the exhaust port. A relatively mildly-tuned engine will have an exhaust port width equal to at least 50-percent of its bore diameter (which is to say, a 3-inch cylinder bore would have a 1.5-inch exhaust port width) and that is enough to allow the ring to spring out into the port window very perceptibly. Make that port square, with sharp comers and sharp edges, and the ring will be destroyed very quickly. And if you enlarge the port so that its width represents 70-percent of cylinder-bore diameter, ring failure would almost certainly occur during the first revolution of the crankshaft. Yet, racing engines have been run quite successfully with 70-percent port widths, and while I cannot recommend that kind of extremity for modified production-type engines, the mere fact that it has been done tells us that techniques exist to make it possible.

Basically, ring-life is improved - at any given port width -by A) rounding the shape of the port window, and B) breaking the window's sharp edges. Both of these measures are employed in all engines, but they reach special and somewhat exaggerated form in racing applications. The traditional port window shape is square, or rectangular, with its corners rounded to help prevent ring snagging. Assuming that the port width does not exceed 60-percent of bore diameter, the radii at the port window's corners should be about 15- or 20-percent of the port width, and that is just what you will find in most engines. But as the port is widened, those corner radii have to be made larger -to about 2s-percent of port width when the latter approaches being 70-percent of cylinder diameter. Actually, even these very large radii will not completely prevent ring snagging if they are not joined by straight-line edges. The upper and lower edges of the port window should be arched, on a radius equal to about twice the port width, in ports having a width that is 60-percent of bore diameter or less. Unfortunately, these simple rules-of-thumb are not adequate at port widths above the 60-percent level- and such widths are - becoming very common. With the rings we had a decade past, which were relatively thick, axially, and quite narrow, radially, (not to mention being made of ordinary and rather brittle cast-iron) the upper limit for port width was 62-percent of cylinder bore diameter; now, with our vastly improved rings it has become possible to widen the port out to 70-percent of bore, or slightly more, *if* we are very careful in shaping the port window.

Just as there is no means of predicting, with any great accuracy, what kind of "cam" and taper a piston will require to fit closely in a cylinder when both are at operating temperature, neither is there any firm rule for shaping ultra-wide exhaust port windows. Both are established, initially, on the basis of past experience, and then modified according to test results. It has been demonstrated, in practice, that a modified

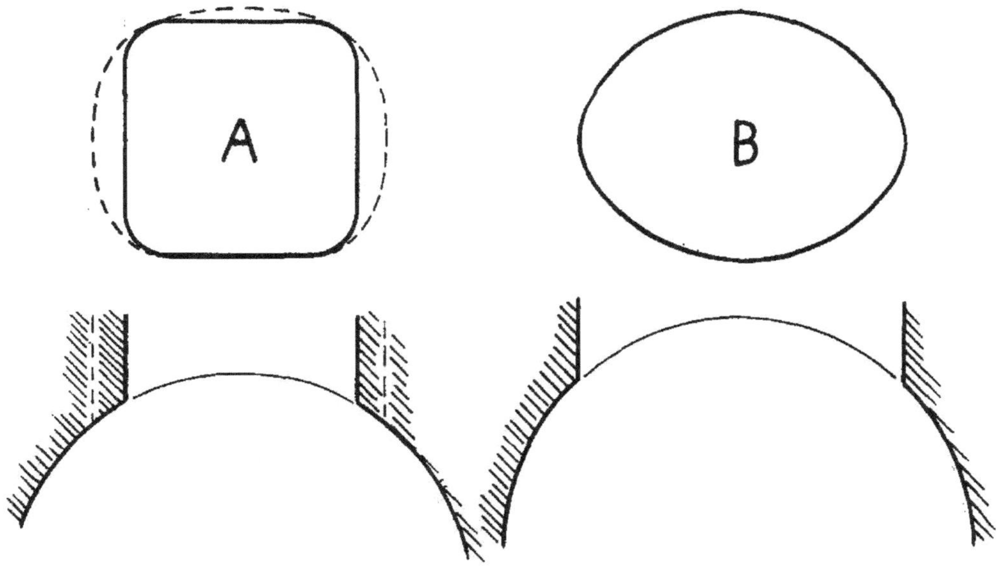

Stock exhaust ports with a width of about 50% of bore (A) may be widened to 62% of bore diameter as indicated by dotted lines. Engines designed specifically for racing have an elliptical, 70% port.

ellipse is the basic shape of port windows in the 62- to 70-percent (of bore) range. Thus, while the ring may actually bulge out into the port window enough to cause its instant destruction in a square port, or in one with straight-line edges are joined with simple radii, the contours of an elliptical port window will sweep the ring gently back into its groove. Then, the only problem that will be encountered is that the ring may bulge out, and be pushed back, unevenly -which may drive one end of the ring sharply against its locating pin and eventually cause the pin to loosen and come adrift. It should be obvious that this last difficulty will be most pronounced when the port window is not perfectly symmetrical, as any departure from symmetry will result in the ring being displaced to one side as it is pushed back into its groove.

Careful craftsmanship will prevent this asymmetrical displacement of the ring; it will not, of itself, forestall other problems associated with very wide exhaust port windows. In my opinion, one should never simply and arbitrarily widen a port out to the 70-percent limit. Instead, the safe and sensible approach is to begin at 62-percent, with a shape that is as nearly an ellipse as is possible. Quite obviously, sharp limitations are going to be imposed by the shape of the existing port window; the idea is to provide the most generous radii permitted by the basic shape with which one must begin. Obviously, too, this reshaping of the exhaust port window will be easier if you have opted for increasing the exhaust timing, as that will give you room to work above the existing port. Then, having established the initial shape, you will have to inspect the rings and the edges of the port window for evidence of scuffing or snagging. Seldom will there be any problem around the lower edge of the port, as the piston slows considerably near the bottom of its stroke. Most of any scuffing that appears will be around the comers of the port; outright snagging will make its presence known in the appearance of scratches leading upward from the center of the port window.

PORT-EDGE CHAMFERS

I have already indicated that the kind of reshaping possible is largely a function of the stock port window's shape, but alterations in shape are not the only cure for scuffing and snagging available to us. Practically everyone knows enough to round-off the port window's sharp edges; what most do not know is that a simple round-off is not what is needed. The purpose in breaking those edges at all is to prevent ring snagging by easing the ring back into its groove, and this job is done best not by a simple radius, but by surrounding the window with a very slight bevel, about 0.080-inch in width, and tapering in toward the window to a depth of perhaps 0.015-inch. It is of course necessary to work a slight radius where the bevel reaches the port window, just to be safe, but the real job of tucking the ring safely away in its groove is performed by the bevel. You will appreciate that the same kind of bevel is needed at all the port windows; those at the transfer ports need not be quite as large.

As regards the exhaust port, a secondary function is served by providing a bevel,

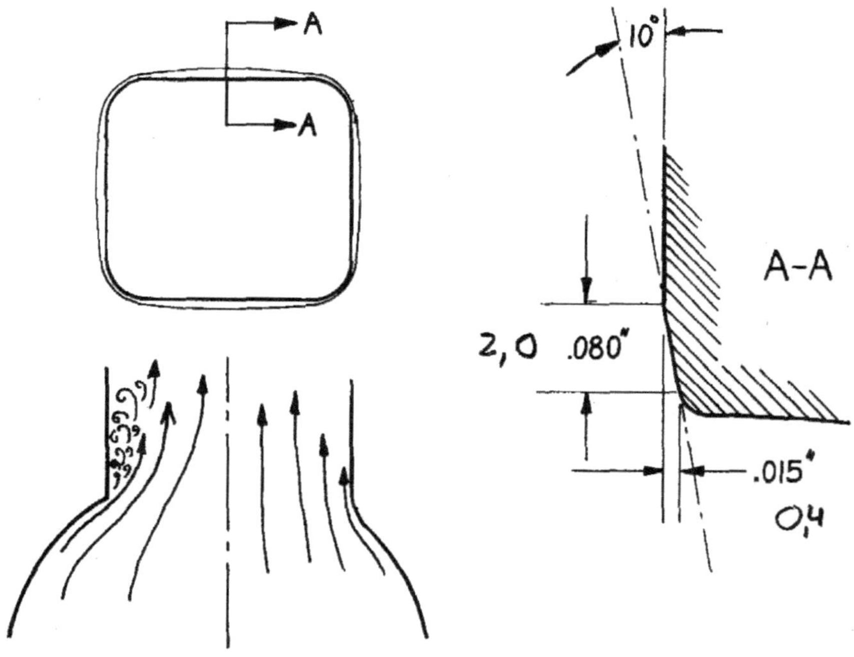

Port window edges should be chamfered as shown in section A-A; a radius at the exhaust port's sides smooths the outflow of gases.

and radiused edges, around the port window. There is a very considerably contraction of flow through any sharp-edged orifice, and such orifices may be made effectively larger by providing them with a rounded entry. Improvements in flow in the order of 30-percent could be had were it possible to give the port window edges a radius of, say, ¼ inch. Unfortunately, to do this would mean advancing the point of exhaust-opening a like amount, which in most engines would result in a very radical exhaust timing indeed. It is, on the other hand, often possible to carve just such a radius at the sides of an exhaust port - although it is questionable that this radius would be as effective as simply widening the port to the same extent. The radius approach does have the advantage of leaving intact much of the metal around the port, which can be important: Thick sections of metal tend to equalize cylinder temperatures and prevent the kind of local distortion that is such a potent cause of piston seizure. Also, in engines having exhaust ports closely flanked by cylinder hold-down studs, there may not be room enough to widen the port as much as would otherwise be desirable, and in that event the side-radiused ports become a necessity.

In the vast majority of engines there will be a fairly large increase in cross-sectional area between the stock exhaust window and the actual exhaust outlet. Indeed, this increase often is too large to give best results with expansion chamber exhaust-systems: What may seem to be no more than a flow-improving enlargement in area leading into the exhaust pipe during the outflow phase of scavenging becomes a sudden constriction for waves returning to the cylinder from the expansion chamber. In fact, if the difference in the areas at the port window and the outer end of the port becomes as great as 1:3, virtually all of the expansion chamber's resonant effects will be lost. What happens, in such cases, is that the waves returning to the cylinder are reflected back into the chamber by the abrupt constriction of the port. Maximum transmission of these waves into the cylinder will, of course, be obtained with a 1:l port-window / port-outlet ratio, but that kind of straight-through passage represents something less than the optimum in minimized flow resistance during the blow-down phase of scavenging (the period beginning when the exhaust port cracks open and ending with the opening of the transfer ports). Thus, the walls of the exhaust port should diverge somewhat, giving a progressively-increasing cross-sectional area out to the exhaust flange. The most important thing to remember, here, is that sudden changes in section should definitely be avoided. Neither gas-flow nor the effects of sonic waves in the exhaust tract are served by a bunch of lumps and jogs - this being far more important than a mirror finish on the port walls. Given an absolutely free hand with an exhaust port (which seldom is possible, when you're modifying an existing engine) I would be inclined to make the exhaust passage's areas equal to those in an 8-degree cone. However, unless you have a dynamometer available for the verification of exhaust passage experiments, the best approach to this whole matter is to leave unaltered the exhaust-flange end of the port, and carefully blend the enlarged exhaust window into the rest of the port.

There is one more item to be covered here: the "bridged" exhaust port. The bridge to which I refer is a vertical rib that divides the exhaust port and creates two separate exhaust windows. Few engines have bridged exhaust ports, but if yours does, it should neither be removed nor narrowed – despite the obvious advantage in gas-flow to be gained thereby. One, and perhaps the most important, reason why we do not see more bridged exhaust ports is that the bridge is inevitably poorly cooled. In consequence, it tends to distort back into the cylinder slightly, and that occurrence becomes the more unfortunate because the piston/cylinder pressure point thus created is very poorly lubricated and localized seizing along the bridge is common. Narrowing the bridge reduces the heat-path into the metal surrounding the port, making the bridge's temperature just that much higher, while weakening it and making serious distortion all the more probable. Despite these considerable shortcomings, the bridged exhaust has been employed and will continue to be in certain applications, for it offers an opportunity to make the exhaust port window area extremely large without the usual attending

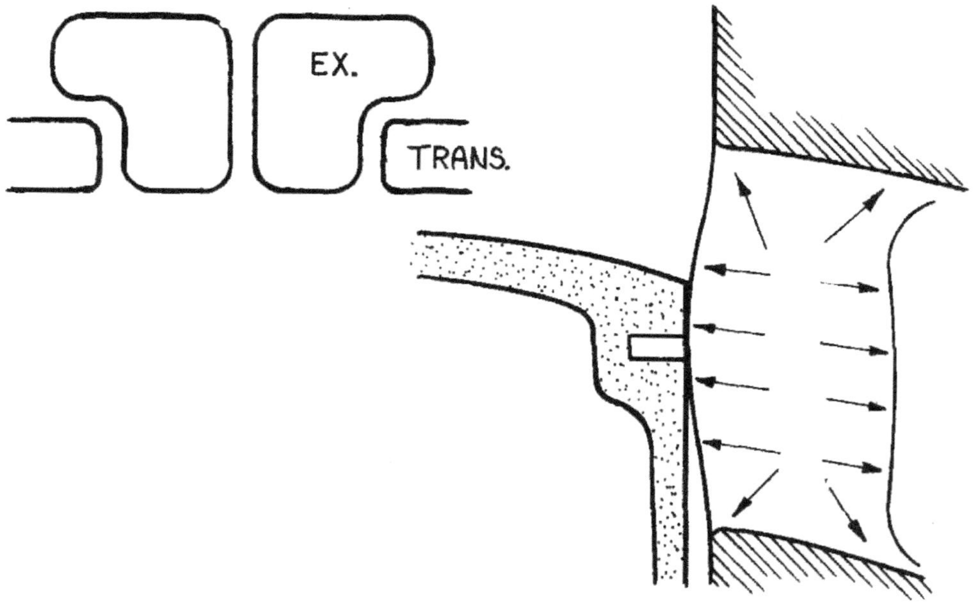

Bridging permist widening the exhaust port until it overhangs the transfers without ring trapping, but the bridge itself may overheat and expand out to create a pressure point on the piston skirt.

difficulties with ring-snagging. On the other hand, it makes worse one of the real problems inherent in oversize exhaust windows -which is piston-crown overheating. At the moment of exhaust port opening, a great blast of fire goes jetting down the side of the piston, and this puts a lot of heat into the piston skirt. Clearly, any enlargement of the window exposes more of the piston to this kind of heating, while depriving the skirt of part of its contact with the relatively cool cylinder wall. Raising the exhaust port height has the same effect, and either of these power-enhancing modifications may, as they are carried progressively forward, require a compensating reduction in compression ratio to protect the piston.

FLOW PATTERNS

Some engines do not respond nearly as well to increased exhaust port width as they should, which brings us to the overall problem of flow in the cylinder during the scavenging operation, and the transfer ports. As was stated earlier, we may imagine the mixture flowing from the transfers and neatly sweeping away residual exhaust products, but it does not really happen in quite that way. For one thing, there simply will not be a volume of gases delivered up from the crankcase sufficient to clear all the exhaust products from the cylinder. Seldom will the delivery ratio (the ratio of piston displacement to air actually pumped into the cylinder) be much better than 1:0.8, or 80-percent. Therefore, a cylinder having a piston displacement of 250cc will only have 200cc of air/fuel mixture coming into it through the transfer port - leaving, by implication, at least 50cc of exhaust gases trapped in the cylinder even if we assume a near-perfect separation of exhaust products and the incoming charge. Actually, there will be some mixing of the two due to turbulence, with the result that some part of the charge is lost out the exhaust port and there is a greater dilution of the fresh charge, with exhaust products, than would be assumed from the delivery ratio alone.

Delivery ratio is almost entirely a function of crankcase pumping efficiency, and the transfer ports' time-area factor - which is to say, the volume of the charge delivered into the cylinder is entirely independent of the number and disposition of the transfer ports. Those things are dealt with elsewhere in this book; we are concerned here with making the most of the mixture actually delivered, and in that regard the importance of the transfers' shapes and placement cannot be exaggerated. Often, the most subtle changes yield very large differences not only in peak power, but in the shape of the entire power curve, and it is all too easy to deal an engine a considerable injury while performing some minor alteration with a steady hand and the best of intentions. In this respect, I think it most unfortunate that the two-stroke engine cannot be driven below a minimum level of operating efficiency by even the most awful butchery of its transfer ports, as an engine thus served will continue to run, and run fairly cleanly, after that kind of surgery. How much better it would be if a serious departure from the optimum would produce a great fit of misfiring, or some other obvious evidence of distress. Unhappily, it will not, which means that an engine's transfer ports should be left strictly alone unless you have both the knowledge and the tools to make any modifications properly. The tools you will have to acquire on your own; knowledge is what I hope to provide with this book.

Perhaps the most valuable bit of information I can supply is that unless you plan to alter fairly radically the speed at which your engine makes its maximum output, there is no need to do anything beyond smoothing the casting flaws out of the transfer ports - and even that should be approached with some caution. Why? Because in scavenging efficiency so very much depends on symmetry of flow. Get one transfer port flowing conspicuously better than its mate on the opposite cylinder wall, and while you may have improved the delivery ratio slightly, the scavenging pattern will have been upset and power output will drop. As a matter of fact, I am inclined to think that the upper reaches of the transfer passages should be left entirely alone, unless to remove some conspicuous casting defect. It is in any case extremely difficult to do precise work up near the port

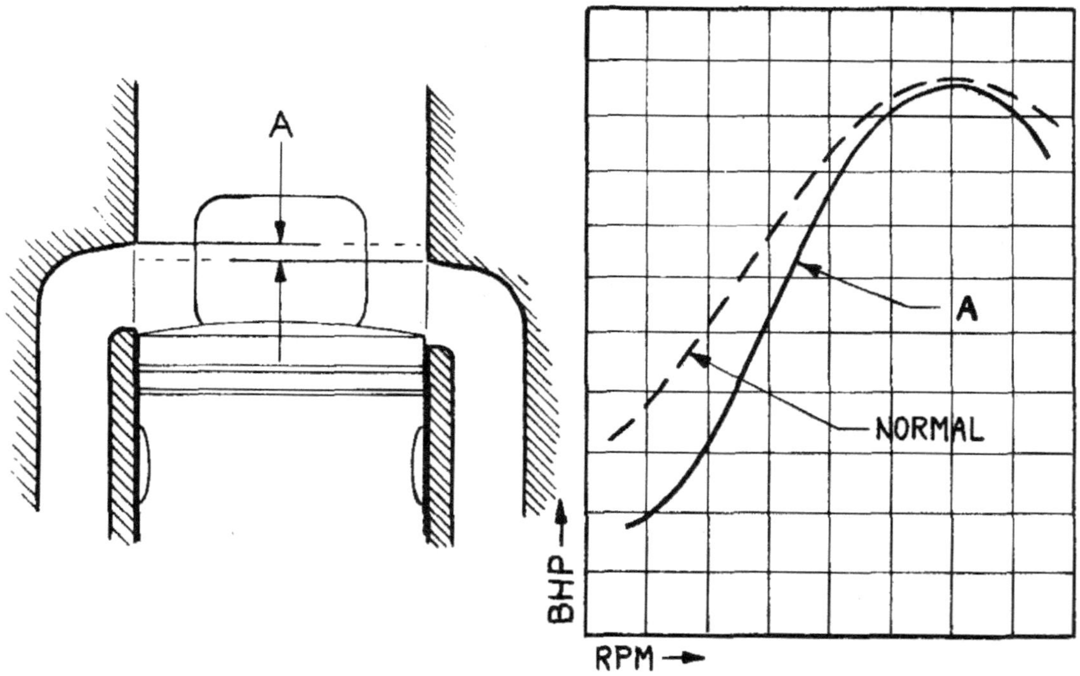

Vertical misalignment of the transfer ports has only a slight effect on peak power; the loss of power is more pronounced at lower speeds.

window, and almost impossible to raise the port's roof without altering its angle - which is the wrong thing to be doing. All things considered, it is probably easiest to raise the transfer ports, when you want to increase the transfer timing, by raising the entire cylinder. A spacer under the cylinder will accomplish this, and it is usually a simple matter to trim the lower edges of the transfers and exhaust port to align with the edge of the piston crown at bottom center. Of course, this method shortens the intake timing, and so you will have to do a bit of trimming there as well, but anything is easy compared with trying to carve higher transfer ports with the port roofs held to their original configuration. *Unless* you discover that lifting the cylinder raises the intake port to the point where the piston-ring ends spring out into it when the piston moves down to the bottom of its stroke, or unless, for some reason, it is not possible to machine a thickness equal to that of your spacer from the top of the cylinder to return to the original compression ratio. When either of those things present a problem, changes in transfer timing should be effected by cutting shallow troughs in the piston crown - which is a measure that can be used on the exhaust side, too, and should be used as a preliminary experiment to see whether the port timing you think you want is what you really need.

All present loop-scavenged motorcycle engines follow Dr. Schneurle's original pattern fairly closely, and many -perhaps most - are more or less exactly the same: Twin

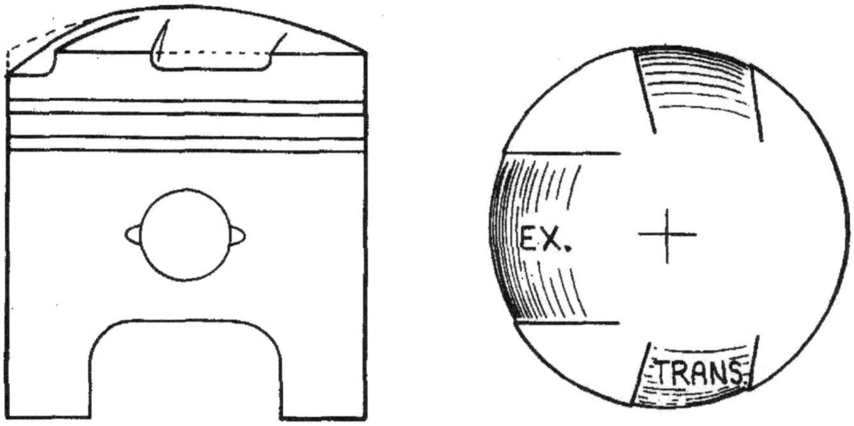

Cutting shallow troughs in the piston crown is an easy, and easily corrected means of experimenting with changes in port timing.

streams of incoming charge emerge from twin transfer ports flanking the exhaust port, and angle back across the piston crown and slightly upward, joining into a single stream at a point approximately two-thirds of the way back from the exhaust port. This stream is deflected upward by the rear cylinder wall, and then it sweeps up to the top of the cylinder to be directed back down the forward cylinder wall - moving the residual exhaust gases out of the exhaust port as it advances in that direction. There is, as you might expect, some turbulence generated by this activity, which is unfortunate because turbulence promotes the very kind of churning and mixing that should be avoided. But the turbulence is minimized when the flow is symmetrical, and there will therefore be less dilution of the fresh charge trapped in the cylinder at exhaust-port closing. Skewing either transfer port to one side, or lifting the upper edge of one slightly higher than the other will badly upset the scavenging pattern

Curiously, the ill-effects of such misalignments are more strongly felt at engine speeds below that where maximum horsepower is produced. Only a slight weakening of an engine's peak output is noted when modifications to its transfer ports leave the scavenging flow badly aimed, but there appears a marked instability at the point of maximum torque and below. This phenomenon would bear investigation, but I suspect that it reflects the fact that there is an almost explosive entry of the scavenging streams into the cylinder at high speeds. Also, it is a matter of record that the modern, short-stroke engine is generally less sensitive to scavenging patterns than the long-stroke engines of years past - this because sheer bulk flow, even if badly directed, does a much better job of scavenging the kind of low, large-diameter cylinder presented by the short-stroke layout than a taller, more slender cylinder. Still, and despite the fact that high horsepower numbers make good conversation, power range is going to be an extremely important consideration until such time as we have transmissions providing infinitely-variable ratios. So the best scavenging system is one that has good bulk-flow capabilities while maintaining a high degree of flow control.

This last item, the proper direction of the scavenging streams, is important for reasons beyond the reduction of turbulence and fresh charge / exhaust products mixing: Cylinders for high-performance engines need very wide ports to avoid excessive timing durations, which means that the ports must be crowded together too closely to entirely avoid the dangers of "short-circuiting" the charge. Having a high delivery ratio avails an engine nothing if the mixture streams emerging from its transfer ports are allowed to divert from their intended path and disappear out the exhaust port. Obviously, this danger increases as the transfer ports are crowded closer to the exhaust port; obviously, too, a degree of crowding is almost inevitable. On the available evidence, it would seem that the point at which short-circuiting becomes a problem is when the separation

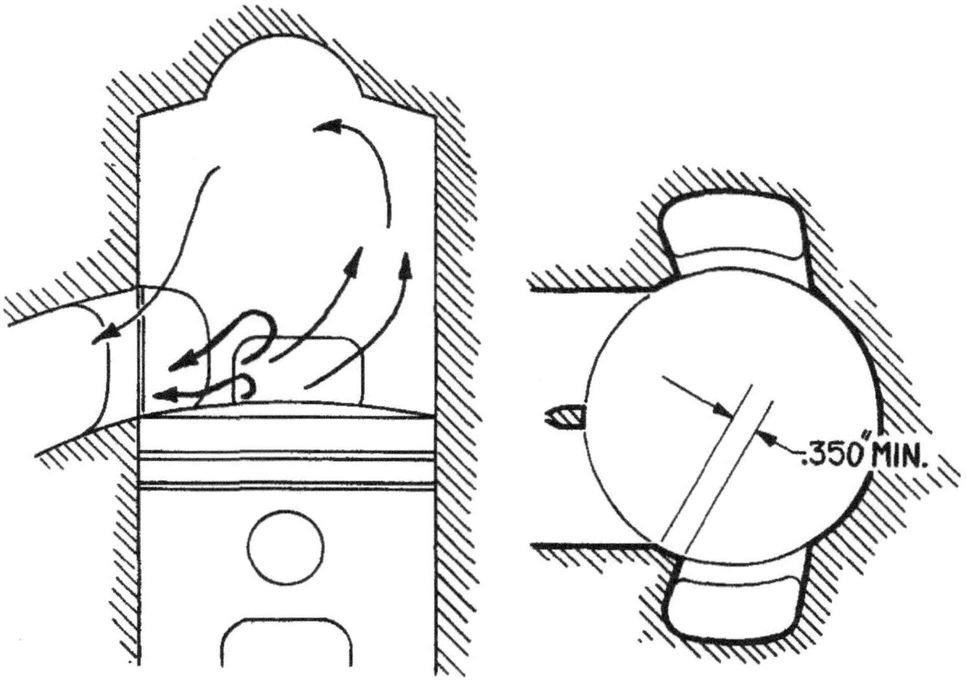

Failure to maintain at least a 0.350-inch separation between exhaust and transfer ports will result in a partial charge short-circuiting.

between the exhaust port's side-wall and the forward edge of the transfer port is decreased below 0.350-inch - that minimum dimension holding true for cylinders of 125cc and up. But it should be understood that this proximity is acceptable only when determined effort has been made to direct the scavenging streams sharply toward the rear cylinder wall.

MULTIPLE TRANSFER PORTS

Some manufacturers, notably Suzuki, continue to employ - and get good results from - a single pair of transfer ports. Nevertheless, the main thrust of development today is toward increasing use of multiple transfer ports. This trend began in the design of engines having other than piston-controlled intake ports, simply because such designs have an uncluttered rear cylinder wall and it was inevitable that someone would add a third transfer port in this otherwise unemployed space. MZ's Kaaden incorporated a third transfer port in his 1962 GP engine, fed from a window high on the piston skirt, just to cool the piston crown and carry oil up to the wristpin bearing. But while improved reliability may have been the motivation behind this change, it also resulted in a power increase. Others found much the same result, and by 1965 the triple-transfer layout was standard for engines having rotary-disc - intake valves.

Isolated examples of piston-port induction combined with multiple transfer ports are to be found throughout the two-stroke engine's history. Still, I am inclined to assign credit for stimulating real interest in the layout to Hermann Meier. Meier was hired by Royal Enfield to design and develop the engine for that firm's short-lived "production racer", but was not allowed anything - approaching complete freedom in this endeavor. Management told him to work around the then-popular Alpha crank assembly (it was a more-reliable replacement for the Villiers crank, rod and bearings) and forbade anything as tricky (and expensive) as rotary-valve induction. In short, he was asked to design a cylinder that would magically produce all the power of a genuine GP engine without any of the latter's complication. Understandably, Meier's best efforts could not overcome the obstacles placed in his path by Royal Enfield's management, but he did develop an engine that was a notch up, in terms of power, over the twin-cylinder Yamaha production racers of that time -which was quite a neat trick. And his Royal Enfield engine did have more than the usual pair of transfer ports. The presence of a large intake port on the rear cylinder wall prevented Meier from positioning a third transfer port there, so he simply crowded the primary transfer ports forward, closer to the exhaust port, and added a pair of smaller, auxiliary transfer ports between the main transfers and the intake port. The auxiliary ports were fed in part directly from the crankcase, but a pair of windows adjacent to and slightly higher than the wristpin, in the piston, also fed mixture through matching holes in the cylinder into the transfer passages.

The 250cc "Grand Prix" Royal Enfield, handicapped as it was by marginal reliability and wretched handling, soon passed into oblivion, but hard on its heels came the now-famous "five-port" Yamaha -with a virtually identical transfer port system. Initially, the Yamaha fed its auxiliary transfer ports entirely through windows in its pistons, and the transfers themselves were merely a pair of grooves flanking the intake ports. This arrangement soon, however, gave way to the now-established Yamaha practice of providing all of the flow for the auxiliary ports from the crankcase; it is only in the recent reed-valve engines that windows once again appear in Yamaha pistons. Without really knowing, I would suspect that Yamaha dropped the windowed piston concept because it adds a complication in manufacturing and weakens the piston - and because Yamaha's needle-roller wristpin bearings survive very nicely the conditions of scanty lubrication and high temperatures that caused problems for Meier and Kaaden.

In any case, multiple transfer ports are much in style, and do appear to give superior results if properly managed. But be advised that simply hacking an extra pair of holes up through the cylinder does not constitute proper management of the multiple transfer concept, for the auxiliary transfer ports must act in concert with the main transfers or they will only contribute to turbulence - which means an increase in charge

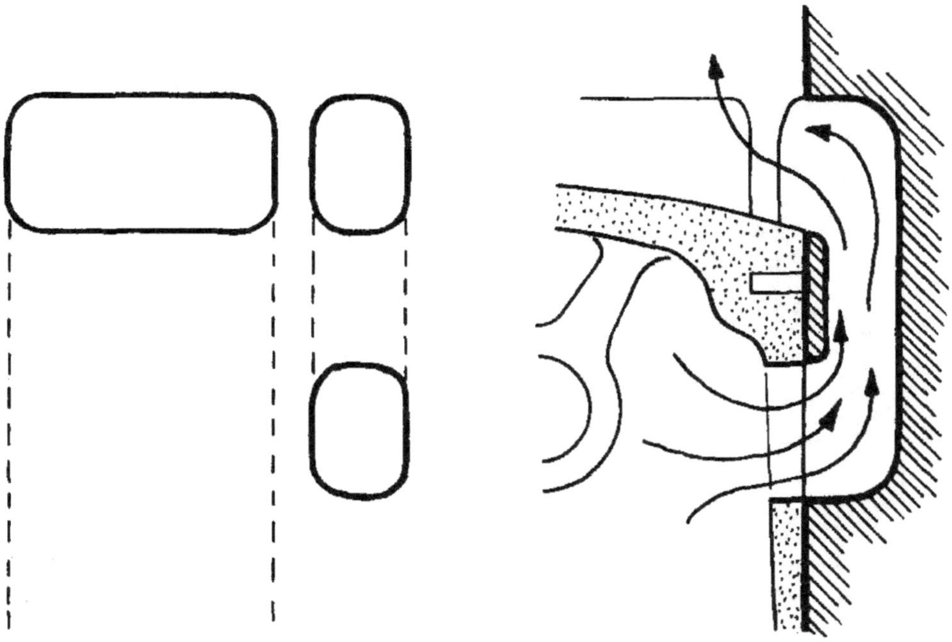

Boost ports fed from holes high in the piston skirt cool and oil the piston crown and wrist pin, while improving power output.

dilution – and / or outright short-circuiting. Unless you have the equipment, and dexterity, to cut extra ports with great accuracy, you would be better off confining your efforts to improving your engine's single pair of transfer ports -if a single pair is what it has.

SUBTLETIES

The best basic shape for a transfer port is a sweeping arc, but many are (for reasons of manufacturing cost or because of space considerations) substantially straight, with a sharp turn where they meet the port window. There is little difference in these two approaches in terms of bulk flow, but scavenging- stream control is superior with transfer passages that sweep wide to join with the port windows less abruptly. Consequently, the peak power obtained will be much the same with either transfer passage configuration, but the less abrupt shape gives better performance below the power peak. Obviously, it is much easier to tell about the difference than to make any real change in the ports' shapes - especially when the engine in question is a twin, with its cylinders crowded together - but

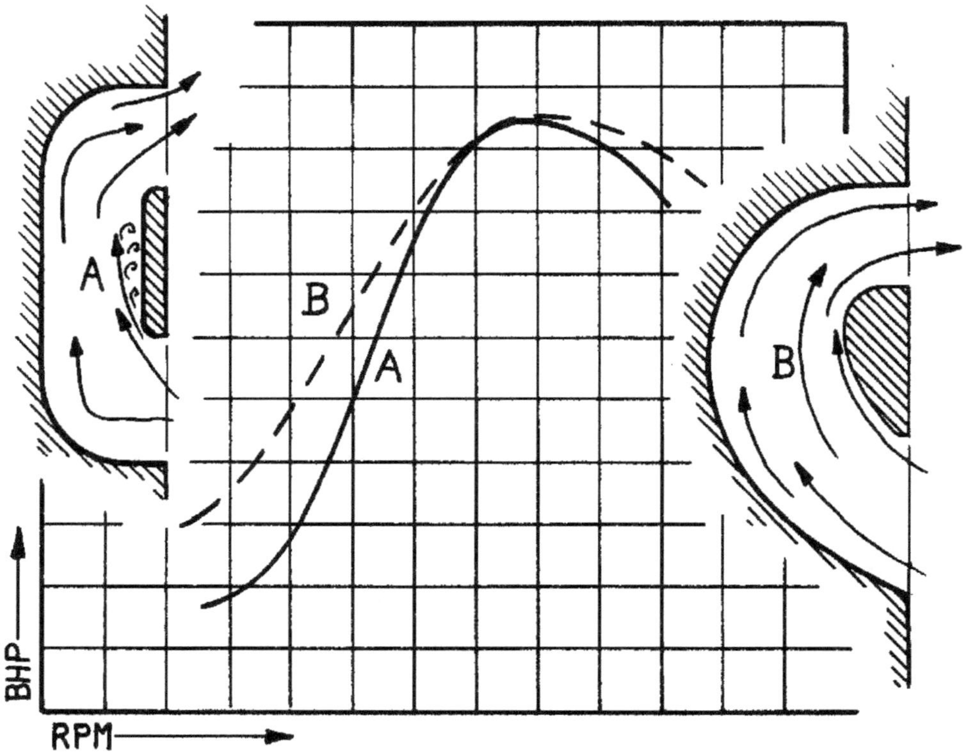

Transfers with abrupt turns (A) match peak outputs of those with sweeping an even arc (B) but are inferior below and beyond the peak.

it also obviously is good practice to do everything possible to make the straight port a little less straight. This can be done by carving away metal from the port's outer wall, and adding a similar thickness of plastic Devcon filler on the wall nearest the cylinder bore. Remember that the transfer passage should taper down from its entrance at the cylinder base to the port window itself, and that it is virtually impossible to make the entrance area too large. Enlarging the transfer passages reduces an engine's primary compression ratio, of course, but we have already established that crankcase compression is not terribly important - and mixture delivery into the cylinder most definitely *is!*

If, or when, you perform the modification just described, be careful not to alter the port window's direction, either upward or to the side. You might improve upon the manufacturer's handiwork in this regard; the chances are that any change will be for the worse. In practice, the effective area of the transfer ports becomes greater when they are angled more toward the middle of the cylinder and less toward the rear cylinder wall, but while a change in the order of 10-degrees will sometimes yield a boost in peak power, this increase will be purchased at some considerable cost in power range and I cannot recommend it to anyone who is unwilling to A) build a flow-pattern test bench and B)

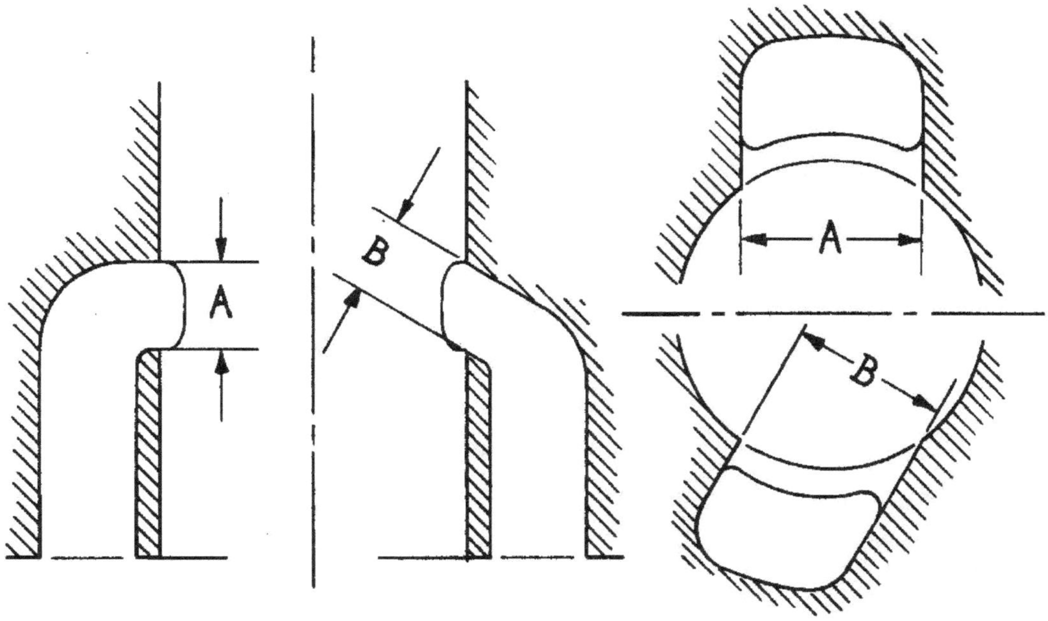

Window areas here are identical, but the effective areas of straight-in ports (A) are greater than those entering the cylinder at an angle (B). Peak power is favored by the former; range by the latter.

dynamometer-test a number of cylinders with incremental realignments of their transfer ports.

Traditionally, the basement-workshop experimenter's scavenging-flow test method has been to feed an engine plenty of oil, and inspect the pattern on the piston crown after a period of running. That is better than nothing, but it provides one with only a very crude picture of what is happening inside the cylinder while the engine is running. Better methods have been developed: One is the “water-analogy” test, in which a cylinder is fitted with a Lucite cylinder head and water forced up through its transfer ports and out the exhaust port. Air is injected into the water, in small bubbles, just below each transfer passage entrance, and these bubbles make the flow patterns inside the cylinder visible. Another, less messy method also entails making a clear Lucite cylinder head, but instead of water being used, air is aspirated from - the exhaust port (with a large vacuum cleaner, if you have one) and smoke or fine dye droplets are injected into the air stream as it enters each transfer port. Only by such means can you actually see asymmetrical scavenging streams, and short-circuiting, and such techniques are essential if you plan any major alteration in the porting your engine's maker has provided.

There are, as it happens, some broad rules concerning the size and direction of multiple transfer ports. One of these is that in a "five-port"" cylinder the auxiliary transfers should have an outlet area between 0.33 and 0.40 that of the main transfer ports. Unfortunately, nothing as clearly defined exists as regards the upward tilt of the transfer ports. In most engines having rotary intake valves and an auxiliary port or ports on the rear cylinder wall, the main transfers will discharge at angles ranging from straight across the cylinder, 90-degrees from the cylinder wall, to as much as 15-degrees of upward tilt. The third port (or ports) will always be tilted upward, but this can range from as little as 45-degrees to as much as 60-degrees. Maximum power is obtained, in most engines, with the least amount of upward tilting - as this makes the port windows effectively larger - but power range tends to suffer as the upward bias in the scavenging stream flow is reduced. Thus, in a motocross engine, one should start with a considerable amount of upward bias and reduce this until the engine's power range verges on becoming too narrow; the opposite approach, that of starting with "flat" ports and adding epoxy filler to the port roofs to provide upward bias, might more profitably be applied in road racing engines. In either instance, changes should be made in one-degree increments until the optimum is found.

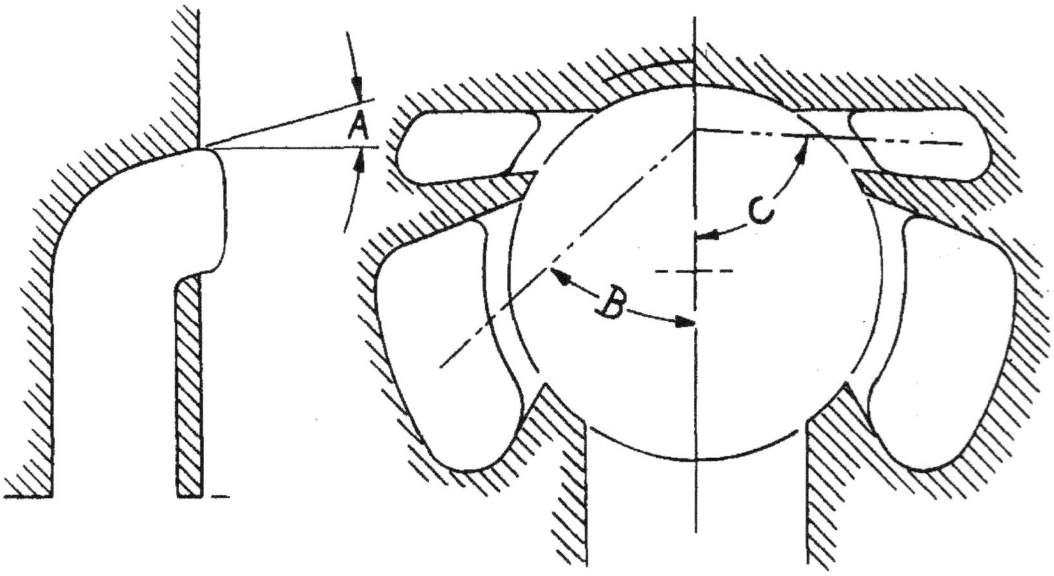

Increases in angle-A, and decreases in angles-B and -C improve the scavenging pattern and increase torque at the expense of peak power, due to the corresponding reduction in effective port area.

There would seem to be no rule at all in determining the upward angles for the kind of paired transfers Yamaha currently employs. The TD2 road racing engine has its main transfer ports angled upward 15-degrees, while the smaller, auxiliary ports discharge straight across the piston crown. But the otherwise very similar, 350cc TR2 engine's main transfers have a zero upward angle, and the auxiliaries angled upward very sharply at 45-degrees. Yamaha's latest road racing engine, the 350cc TR3, combines these, with its main transfers directed upward about 15-degrees and the auxiliaries even steeper at about 45-degrees. In all cases, both main and auxiliary open at the same time in the Yamaha engines, although this is not universal by any means. Some designers have obtained best results by either advancing or delaying slightly the opening of the auxiliary transfers, with respect to the main transfer port timing, and the old GP Suzuki 50cc single had its third port open before its exhaust port - which caused crankcase pressure to rise very sharply just before the exhaust blow-down and subsequent transfer phase. Of course, this rather radical method of improving mixture delivery was effective only over an extremely narrow power range, and even then it must have had other serious shortcomings for neither Suzuki nor anyone else has employed it in recent years.

Now that reed intake valving has come into fashion, there will be a lot of "sixth and seventh" porting done to various engines. There are two different approaches to adding extra rear-cylinder ports in reed-valve engines: That chosen by Yamaha is simply to carve a slot leading up from the intake port to link the port cavity with the cylinder when the piston is down at the bottom of its stroke. This "sixth" port is not fed from the crankcase; mixture moves up through it and into the cylinder due to the ramming action of gas inertia in the intake manifold, and also because of the partial vacuum formed in the cylinder by exhaust system dynamics. Yamaha places the top of this extra transfer ports on a level with the four already found in their five-port engines, and the roof of the port is angled upward very sharply - about 70-degrees, in fact. Apparently, such "tuning" as is done with this kind of port is mostly a matter of varying its width, but not enough experimental work has been done with it to provide us with any guidelines. The other approach to this particular kind of transfer port is to cut windows in the rear of the piston to permit part of the charge compressed in the crankcase to blow up into the intake port, from where it is then able to escape through the boost port and into the cylinder. From early reports, the effect of this modification is to improve very substantially an engine's power range at some expense to peak power. This is, of course, precisely the effect to be desired in a large-displacement scrambler or motocross engine, but hardly what one would be looking for from a road racing engine. My own experience with reed valves is still too limited to permit me any firm conclusions, and the potential benefits of the porting they make possible, but I am inclined at present to believe that they fall

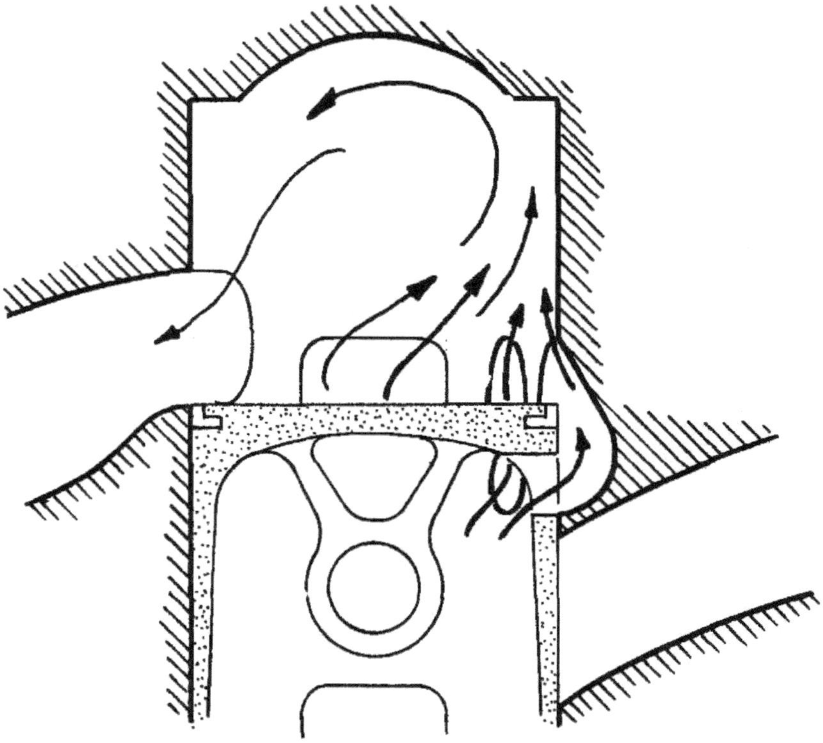

A promising approach to boost porting is this one being tried by Aermacchi in conjunction with a flat-top piston and dykes ring. Triple boost ports are fed through holes high on the piston.

something short of being the answer to the two-stroke tuner's prayers. Indeed, it may be that their principal contribution is to keep carburetion clean over a wider speed range than is possible with piston-controlled intake porting. I very much doubt that reed valving will ever equal the ordinary piston-controlled induction in terms of maximum horsepower - though I recognize that reeds have their uses in engines intended for general, all-purpose motorcycles.

Far more promising as a mean of completing the ring of ports around the base of the two-stroke's cylinder is a method being tried by Aermacchi, in Italy. Engineers these have been experimenting with a taller-than-usual piston, which provides them with enough space above the intake port for a trio of very short transfer passages fed from three holes drilled through the back of the piston, just below the ring. To get room for these holes, without pushing the piston crown up absurdly high above the wristpin, Aermacchi's engineers have gone to a flat-crown piston, with a single Dykes-pattern ring located at its upper edge. A similar modification might be possible in other piston-port engines, without getting into too much special machining or prohibitive costs.

CYLINDER SCAVENGING

No matter what kind of transfer system you ultimately employ, remember that while adding port area will improve bulk flow of the air / fuel charge up from the crankcase, it will all be for nothing unless the flow is controlled after it enters the cylinder. In a cylinder with main and auxiliary transfer ports, the upward pitch of the auxiliary ports is used to literally *herd* the main scavenging streams together and direct them up into the cylinder. This can be accomplished by shooting the auxiliary streams under the main flow, and bending it upward - or by aiming the auxiliary streams upward and using them to drag the main flow along. The latter is to be favored when it is possible, simply because - as I have stated repeatedly - tilting a port upward reduces its effective area and reducing the area of the auxiliary ports has less effect on the total flow characteristics. On the other hand, if your engine has its existing transfer ports angled upward at, say, 15-degrees, and is constructed so that only very limited space is available for an extra pair of ports, then it might be better to aim the auxiliaries straight across the piston crown - as was done by Yamaha in its early experiments with the five-port concept.

CARBURETION, IGNITION

No greater futility exists than to invest heavily of one's time and money in building an engine with mirror-smooth ports, the best available pistons, rings, etc., and then approach the problems of carburetion and ignition armed only with enthusiasm. It is an observable fact that some inherently mediocre engines have been made to prevail in racing simply because the men charged with selecting jetting and spark timing knew their business, but never in the history of the sport has there been an engine so splendid in its internal arrangement that a tuner's ineptitude could not render it entirely ineffectual. As a rule, two-stroke engines are both more difficult to tune and more sensitive to fine-tuning than their poppet-valve cousins - especially as regards spark timing, with some two-stroke engines being so extraordinarily sensitive to ignition advance that a shift of only a single degree either side of the optimum will result in a measurable loss of power. Yamaha's TD-series racing engines, for example, must be timed plus or minus only 0.0012-inch of piston travel from the specified distance of TDC. These engines are also very sensitive to spark intensity, and magneto maladjustments or other defects too slight to produce perceptible misfiring will cause a marked deterioration of performance. The two-stroke engine's air / fuel mixture is a somewhat less touchy matter, as over-rich mixtures tend to compensate for their reduced effectiveness in burning by improving internal cooling, but there is an optimum here too. Finding these optima for spark and mixture can be an exasperating, time-consuming chore, but there is no escaping the necessity for putting forth the effort.

While most tuners would agree that much depends on the basic tuners' arts, they are nonetheless inclined to seek magical solutions to any problems not instantly resolved by changing a main jet or replacing a set of points. Sadly, seldom does mechanical exotica - magic-box ignitions and the like - truly solve a problem. More often they merely replace the existing difficulty with another, or multiple, even more mysterious problems. For instance, most engines have ignition systems entirely capable of producing sparks at a rate adequate for the speed range anticipated by the engines' makers, but may collapse into a fit of misfiring at higher revs. Most tuners will assume that the root cause of this distress is that they have built themselves such an incredibly vigorous engine that nothing weaker than God's-Own-Lightning is enough to make it run properly, which sends them scurrying away in search of some transistorized, magnetically-triggered system with enough sheer zap to fracture atoms. Often as not, the system they get will clear up the misfire and convince them that they have done a Good Thing-even though the new system is impossible to time ,accurately and is prone to sudden, inexplicable failures. Actually, the fundamental reason for the misfiring probably was that at the higher speeds, the point cam profile was too “sudden” for the breaker-point springs, leading to point-float. The proper cure would have been to modify the cam profile to open and close the points more gently, and / or substitute the standard points-set with one having stronger springs, instead of abandoning the straightforward mechanical problem in favor of a black box of troubles best left to electronics engineers.

Details vary according to the fashion of the moment, but much the same series of events often occur subsequent to the discovery of a carburetion problem. A persistent misfire frequently leads people into the error of assuming that the carburetor at hand must be replaced with something more splendid. Today, the most frequently chosen replacement is a Mikuni carburetor, and for once the dictates of fashion are substantially correct, as the Mikuni is a very fine instrument and when used in the appropriate throat size it can be made to give excellent results. Unfortunately, it is distinctly possible to employ even the Mikuni carburetor badly, unless you know what you're doing and are prepared to work at it diligently.

THE BASIC CARBURETOR

There is nothing quite so simple, and at the same time complicated, as a carburetor. Your dictionary will tell you, correctly, that it is merely a device for mixing fuel (in this context, gasoline) with air, and although early examples employed several different means to this end it has long since been proven that the job is handled best by directing the air through a venturi in which a fuel-entry nozzle is incorporated. Pressure within the venturi is reduced below atmospheric in direct proportion to air velocity, which applies a suction to the fuel nozzle in the same proportion. Thus, fuel is drawn from the nozzle at a rate closely tied to air velocity, and the basic proportions of the combustible air / fuel mixture produced by the carburetor may be established by varying the cross-sectional area of the fuel entry as compared to that of the venturi. And, in point of fact, early carburetors were little more complicated than the device just described, with only a controlled-level fuel reservoir (the float chamber) and some form of throttle valve added.

Modern carburetors are still built around the basic venturi, fuel nozzle, reservoir, and throttle, but various details have been added. The most important of these is the "correction-air" feature, which is needed to compensate for the venturi / fuel-nozzle's great inherent defect: Uncompensated, the basic carburetor will deliver an air / fuel mixture in which even higher proportions of fuel appear as air velocity through the venturi increases. The reason for this is that the pressure-drop within the venturi is accompanied by a reduction in air density, and while flow from the fuel nozzle is in direct proportion with air velocity, the actual mass of air passing the nozzle *does not* remain in proportion. In consequence, mixture strength rises with increases in velocity unless measures are taken to prevent that from happening.

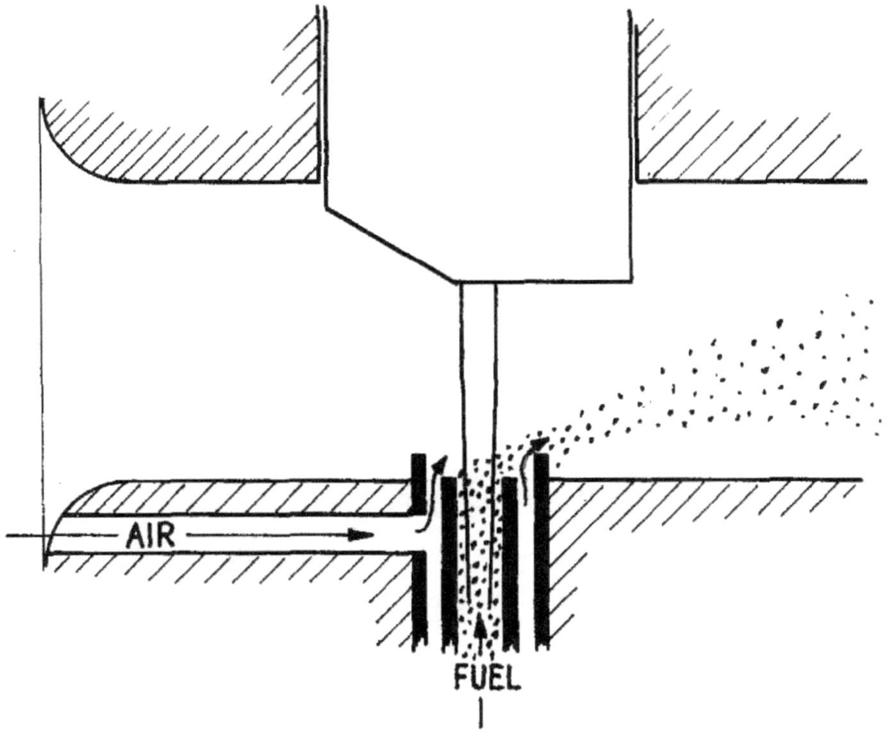

Air entering from outside the carburetor throat compensates for the fact that the depression in the throat increases faster than bulk air flow and prevents mixture strength from increasing with flow.

All carburetors except those with diaphragm-controlled metering make the necessary adjustment for decreasing air-density by means of a correction-air system, in which air at atmospheric pressure is delivered to the fuel nozzle via drilled or cast passages. In its more primitive forms, the air is simply led to an annular opening around the spray nozzle - as may be observed in Amal carburetors. As pressure within the carburetor throat drops, the differences in viscosity cause air-flow from the correction-air system to rise more rapidly than fuel-flow from the spray nozzle, and this does much to stabilize mixture strength. But much better mixture stability is obtained by directing the correction-air into a well below the spray-nozzle, from which it is pulled through a series of holes in an "emulsion" tube. This emulsion tube, which may be incorporated as part of the spray nozzle, takes in fuel at its lower end (usually) and admits the correction-air through a pattern of holes drilled in its sides. At low rates of flow, the well is almost entirely filled with fuel, and air passes only through the topmost holes in the emulsion tube. But as overall flow increases, the fuel level in the well drops to uncover more holes and the correction-air component of the froth delivered to the spray nozzle becomes larger.

By altering the hole-pattern in the emulsion tube, and with adjustments to the total amount of air being admitted to the correction-air system, the carburetor's mixture-delivery characteristics are tailored to suit a particular engine. It must be noted here that an absolutely-even mixture strength may not be desirable: Some high-output engines require a richer mixture at their torque and / or power peaks than at other speeds, which means that the correction-air system should be adjusted accordingly. Generally, a large main jet and small correction-air jet deliver a mixture increasing in richness with engine speed, (as you might expect), with the opposite being true of small main jets and large correction-air jets. But while the overall slope of mixture delivery is determined by the main jet / correction-air jet proportions, slight periods of richness and leanness may be created by alterations in the emulsion tube's hole-pattern. For example, an emulsion tube with large holes at its upper extremity and smaller ones farther down will tend to deliver a richer mixture at higher revs; the opposite condition is found when the upper holes are smaller than those in the rest of the hole-pattern. And, when the emulsion tube is a close fit inside the fuel well, it is possible to make a further adjustment with differences in diameter down the length of the tube: a collar, midway down the emulsion tube, can constitute a restriction within the fuel well large enough to become a kind of secondary

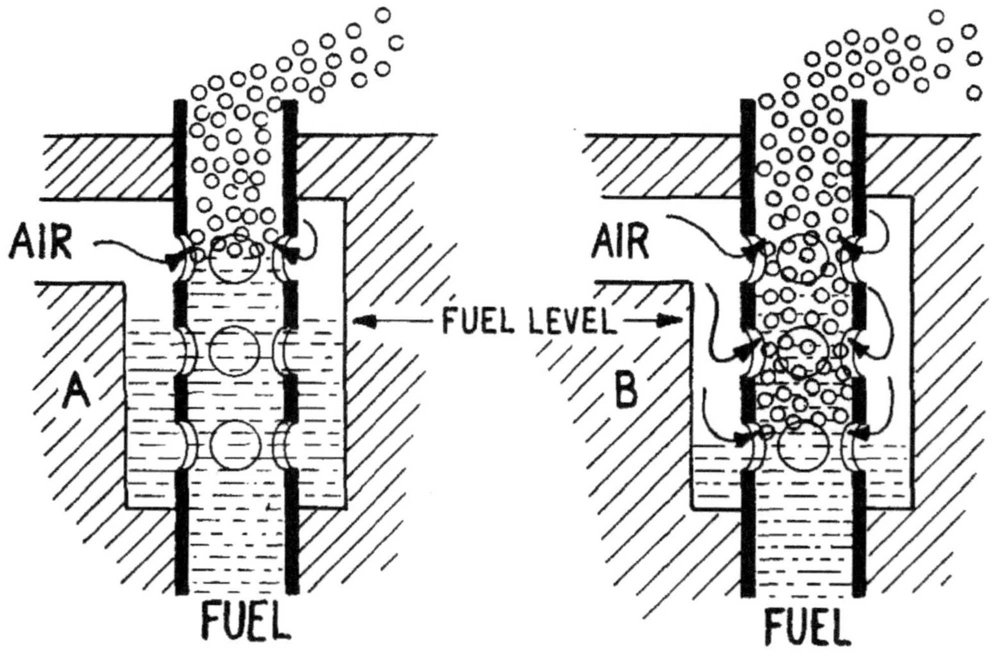

In some carburetors compensating air is fed around an emulsion tube, in which the fuel level drops as throat velocity increases and brings extra sets of bleed holes into action.

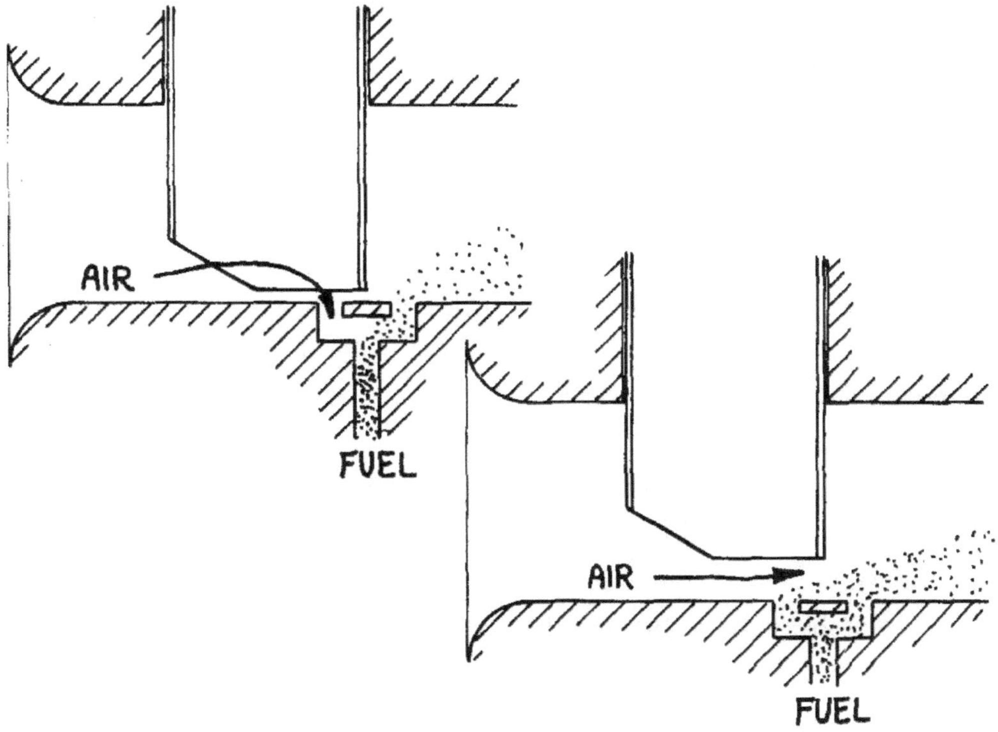

A refinement in idle mixture delivery is effected by having two fuel feed holes in the carburetor throat. The upstream hole functions as an air bleed at idle but switches to a fuel jet as the slide lifts.

correction-air jet. With maximum application of these mixture compensating techniques, it becomes possible to employ very large carburetor throat sizes relative to cylinder displacement, which is why the sophisticated Mikuni is a better choice than the Amal GP-pattern carburetor despite the latter's unquestioned advantage in air-flow capacity, size for size. A 35mm Amal GP will flow more air than a 35mm Mikuni, but you can fit a 40mm Mikuni on an engine that would develop the blind-staggers with an Amal GP larger than 35mm in throat size.

Other complications in carburetor construction have been introduced to cope with part-throttle conditions. At very low engine speeds, for instance, air velocity through the carburetor will not be high enough to pick up fuel from the main spray nozzle and the mixture must be provided by other means. Typically, idle-speed mixtures will be produced by a kind of carburetor within the carburetor: Under the floor of the carburetor throat, where the throttle valve seats, you will find a small chamber supplied with fuel and air, and with entry to the throat through a single hole downstream from the throttle valve, or through one hole downstream and one or more upstream from the throttle. The least complicated arrangement is one in which there is only a single hole behind the throttle valve, and the chamber is fed air past an adjustable needle valve while fuel is supplied through an orifice of fixed size. Here, the idle-speed mixture (which is a froth emerging from the entry hole in the carburetor throat) is controlled by varying the amount of air admitted to the chamber, and the idle mixture will be full-rich when the needle valve is closed. Other systems have a fixed pilot-air jet and an adjustable fuel jet, which reverses the rich / lean position of the needle - while still others have fixed air *and* fuel jets, and bulk flow from the idle-mixture chamber is controlled by the adjustable needle jet. These details are relatively unimportant to the tuner; what is important is that the idle-mixture system not only keeps the engine running at low speeds but also handles the transition between closed-throttle running and the point at which enough air is flowing through the throat to initiate the movement of fuel up from the main spray nozzle.

The transitional period is managed best by having feed holes both in front of and behind the throttle valve. With the throttle against its stop, some air passes under the valve and picks up the mixed fuel and bubbles coming from the idle feed hole, while air is diverted down through the hole in front of the throttle valve to mix with the fuel in the idle-mixture chamber. But as the throttle opens, the depression existing behind the valve moves forward to cover the upstream feed hole, which means that instead of air entering the hole, fuel is pulled from it, and the net result is that the carburetor is delivering enough additional fuel to compensate for the increase in air moving past the opening throttle. That is, of course, only the case when all the fuel and air passages are the right size - and when they *are,* the mixture will remain at the proper proportions until enough velocity is established past the main fuel nozzle to relieve the idle system of further duties.

Carburetors with butterfly-type throttles often have a pattern of holes ahead of the throttle valve, and these are called "progression" holes. As the valve disc pivots, and its lower edge swings forward, it moves ahead of the progression holes in order of placement, and each hole then switches from being an air-bleed to become a fuel jet. With the right pattern of progression holes, even a very large carburetor (large in terms of throat size relative to cylinder displacement) can be made to keep the engine running without stumbling while the transition is made to fuel flow from the main nozzle. Rarely are progression holes, in the multiple, found in slide-type carburetors. In these, the transitional period is handled by the slide cutaway - and the higher the cutaway, the leaner the transitional period's mixture.

ADJUSTING MIXTURE

Almost certainly, the carburetor you will be using will have a circular-slide throttle, because this is the type most commonly employed, and most successful, in the

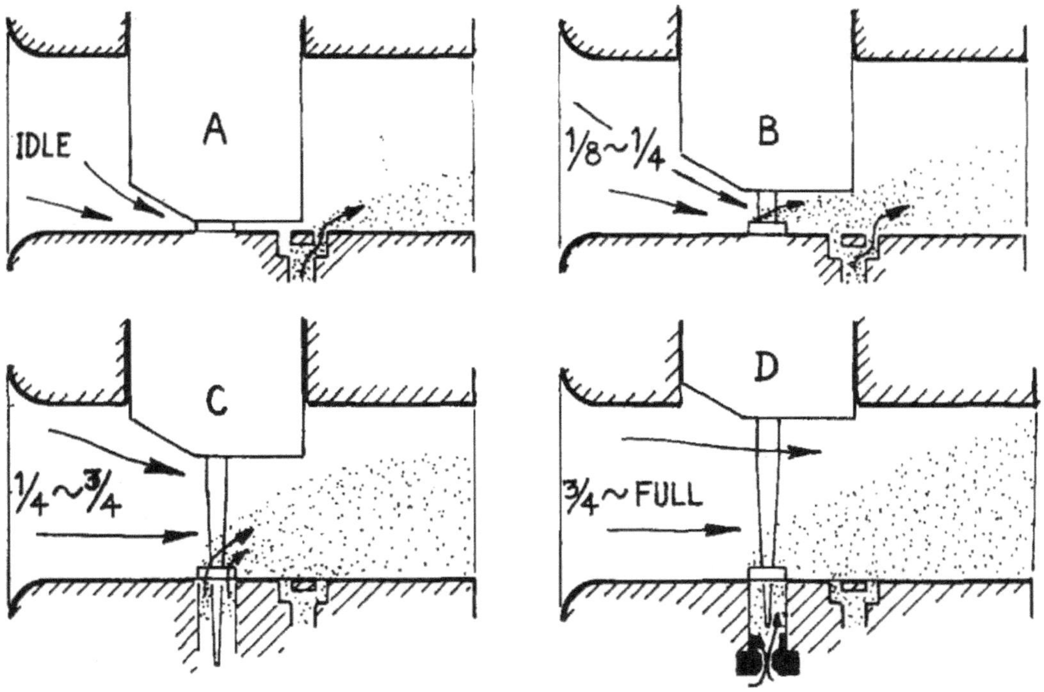

Idle mixture is supplied solely by the idle circuit (A). The action of the throttle slide cutaway helps pull fuel from the main nozzle (B) but its influence fades above ¼ throttle and metering is accomplished by the needle and needle jet from ¼- to ¾-throttle (C) after which the main jet takes over from ¾ to full throttle.

field of high-output motorcycle engines. Almost certainly, too, the carburetor you buy for your racing engine will have been jetted and given a slide cutaway suitable for a somewhat larger stock engine, which should warn you that a more-or-less complete retuning of the instrument will be necessary. Many tuners begin the retuning process by finding the correct main jet, and that is a good beginning unless there is an undiscovered problem with the mid-range metering system - the long, tapering needle clipped to the throttle slide, and the needle jet itself. These, in combination, constitute a variable fuel-metering valve, and if the flow is restricted between the needle and needle jet to an extent greater than the restriction provided by the main jet required to feed the engine at full throttle, then no amount of switching main jets will get the engine running properly. So the retuning process must always begin by determining if there is sufficient fuel flow past the needle jet to feed the engine. I have found that this matter can be settled very simply by lowering the needle to its last notch, which maximizes the flow restriction at the needle jet, and then removing the main jet entirely from the carburetor. The engine should then run, if none-too-well, on part-throttle, but flood as the throttle is opened fully. Should the engine be willing to run on full throttle, you may be sure that a larger needle jet is required.

After the selection of a needle / needle jet combination that will pass more fuel than the engine can digest, you then proceed to the problem of finding the correct main jet. Until you become really expert in the art of "reading" spark plugs, the right approach is to start with a huge main jet and then reduce the size until the engine will just barely run, on full throttle, without "four-stroking". Because of the benefits in cooling that are obtained with very rich mixtures, you will get very near the maximum power to be had from a high-output two-stroke engine with a mixture that verges on being so rich that misfiring occurs. The optimum usually will be found with a slightly leaner mixture than that bordering on four-stroking, but as the potential gain is rather small, and the risk of melting a piston is very large, leaner mixtures should be tried very cautiously.

While you are experimenting with main jets, the metering needle –which controls the mixture strength when the throttle is from about a quarter to three-quarters open - should be installed with its clip in the middle groove, or halfway through its adjustment range. You ultimately will probably change this setting, and perhaps switch to a different needle, but you'll need a starting point for the adjustments required to the throttle cutaway and the idle system. Start these adjustments by backing out the throttle-stop screw until the throttle is completely closed, and then turn it back in until the throttle is just barely cracked open. Having done that, you also close the idle mixture screw completely, and then open it two or three turns before starting the engine. The object, in juggling the two adjustments, is to keep working with the idle-mixture adjustment to increase the idle speed while dropping the idle by backing off on the throttle stop. Eventually you'll arrive at the lowest throttle setting at which the engine will idle satisfactorily, and the correct mixture at that throttle opening, unless, of course, the idle system is wrongly jetted. On carburetors with an "air" adjustment, and a fixed jet feeding fuel, you'll know that the fixed jet is too small if engine speed continues to rise (at a fixed throttle-stop setting) until the adjustment screw is turned all the way in to the closed position; at some point in this process the idle mixture should become over-rich, and if it does not the fuel jet is too small.

The opposite is, of course, true when engine speed continues to climb as the idle mixture adjustment screw is opened, without an optimum ever appearing. A similar, but opposite, rule applies for idle systems with fixed air jets and an adjustment for fuel flow. I cannot give you a listing of which manufacturers use what type of idle system, as this varies even between carburetors of a single make. To be certain, you'll have to take your carburetor apart (or check the maker's literature, if available) to see whether the mixture adjustment is for fuel or air.

Throttle cutaway will be the next matter for your attention, and this factor almost exclusively concerns what happens during the first eighth of throttle opening. It is

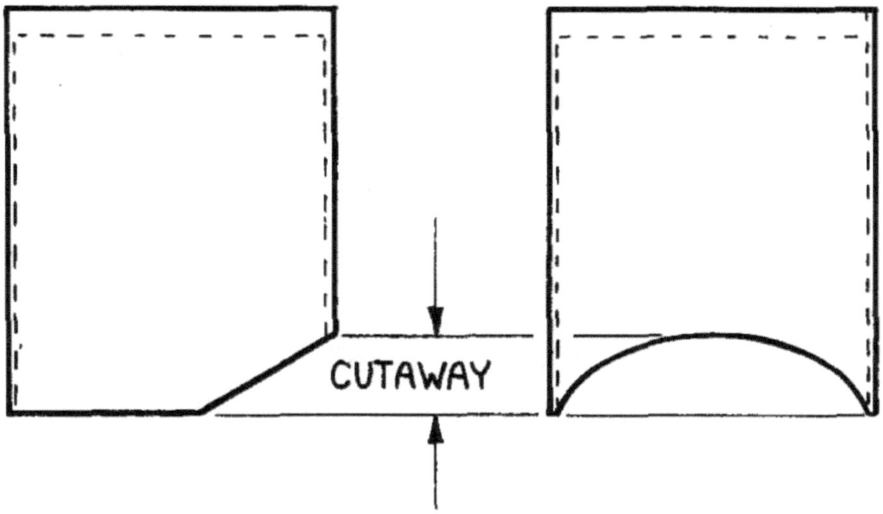

Throttle slides are numbered according to cutaway height, with the higher numbers indicating greater cutaways.

possible to have a too-low cutaway on the front edge of the throttle slide, but a carburetor intended for some big engine will almost always have too much cutaway for one with a smaller cylinder displacement. The too-high cutaway problem will be manifested in a tendency for the engine to cough and die when the throttle is opened, and the cure simply is a new throttle slide with less cut-away. When dealing with a single-cylinder engine, you can buy a replacement slide with the lowest cutaway offered, and file the bevel higher until off-idle running is clean. The same approach may also be taken with multiple cylinders, but it is very difficult to get the cutaways modified exactly the same and if finances permit you should just buy a selection of slides. Incidentally, a too-low cutaway will make the engine surge and burble at one-eighth throttle or less -and if you are forced to make a very large change in cutaway, you'll have to start the idle-system tuning process over from the beginning. Probably the best sequence in overall carburetor tuning is to begin with the main jet, then rough-adjust the needle and needle jet, after which the cutaway and idle jetting are managed more or less simultaneously as it is almost impossible to separate them completely.

Factories devote months to finding precisely the right metering needle, because a touring motorcycle spends most of its life being run somewhere between one- and three-quarter throttle. Fortunately, considerations of fuel economy are strictly secondary in racing, so you need not spend months switching needles and needle jets, but a racing motorcycle is much easier to ride when it at least runs cleanly on part throttle and you

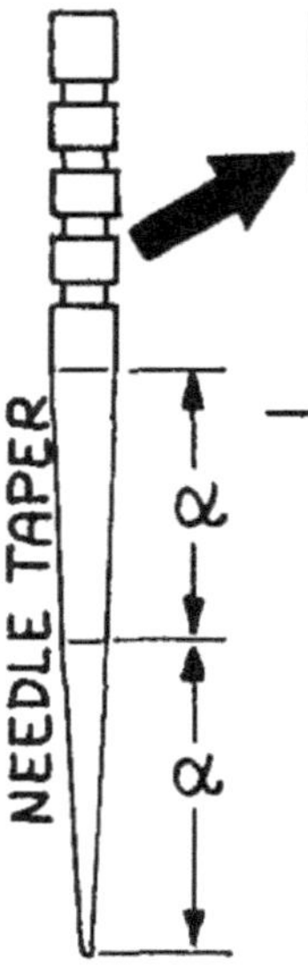

A=0°15'	B=0°30'	C=0°45'	D=1°0'	E=1°15'	F=1°30'	G=1°45'	H=2°0'	I=2°15'
J=2°30'	K=2°45'	L=3°0'	M=3°15'	N=3°30'	O=3°45'	P=4°0'	Q=4°15'	R=4°30'
S=4°45'	T=5°0'	U=5°15'	V=5°30'	W=5°45'	X=6°0'	Y=6°15'	Z=6°30'	

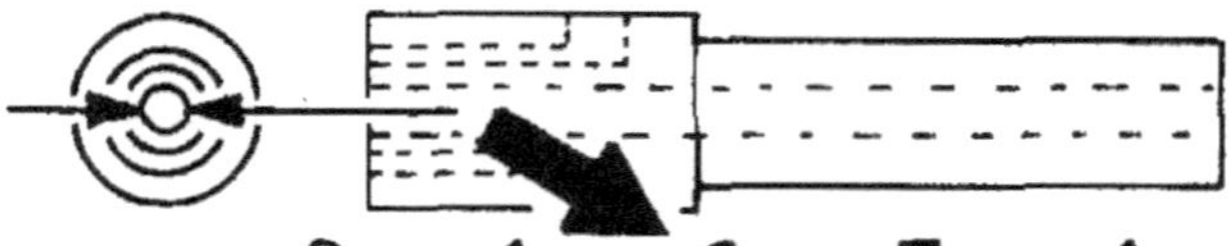

NEEDLE JET DIA.

	0	1	2	3	4	5	6	7	
N	2.550	2.555	2.560	2.565	2.570	2.575	2.580	2.585	etc.
O	2.600	2.605	2.610	2.615	2.620	2.625	2.630	2.635	"
P	2.650	2.655	2.660	2.665	2.670	2.675	2.680	2.685	"
Q	2.700	2.705	2.710	2.715	2.720	2.725	2.730	2.735	"

Mikuni needles are double tapered, and the two letters in their identification numbers are a code for the upper and lower tapers, respectively. Needle jets have a letter and number identification indicating their inside diameter, as shown in the lower of the charts presented.

will have to make some effort in this direction. If there is surging and stuttering at steady throttle within the range controlled by the needle, then the mixture is too rich and the needle should be lowered. An engine that runs fairly cleanly at steady throttle but stumbles and hesitates as the throttle is opened farther is suffering from mid-range leanness and the carburetor's needle should be raised. Sometimes you will get both symptoms with the same needle, with an over-lean condition at one-quarter throttle changing to become over-rich as you approach three-quarters throttle. That should tell you the needle's taper is wrong, being too steep, which means that n needle with a more-shallow taper will be required. Obviously, the opposite may also be encountered. Unhappily, these needles are expensive, but there is no satisfactory alternative to buying a selection and trying them until the right one is found.

At one time a remote-mounted float chamber was the *sine qua non* of racing carburetors. Commonly, carburetors were bolted rigidly to the engines they fed and float chambers were mounted in rubber to isolate them from the worst effects of engine vibration - which can cause such extreme frothing of the fuel that accurate metering is impossible. But while remote mounting may have solved one problem, it introduced another: Acceleration and braking caused a surging in the long fuel lines linking the carburetor and float chamber which also upset metering. The answer, all along, was to incorporate everything in one casting and mount the whole carburetor/float chamber assembly in rubber. The difficulty in that was finding a rubber capable of withstanding a

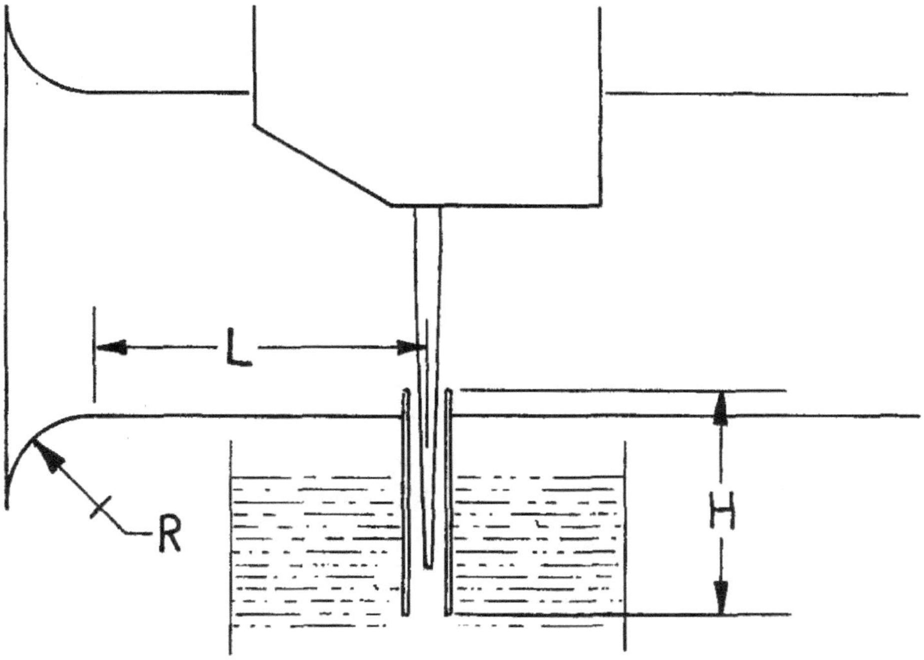

Distance "L" should be from 1.5 to 1.6 throat diameters. Needle jet length provides a column of fuel that damps out the influence of pulsations in the carburetor throat that would otherwise upset fuel delivery; "H" should be at least 12mm. The radius of the carburetor bell should simply be as large as possible within available space.

lot of heat, and constant exposure to gasoline, without disintegrating - and a rubber that could be bonded to a steel manifold flange to avoid the inconvenient space requirements of the old hose-and-clamps assembly. Advances in synthetic rubbers eventually brought us the right material, with the result that nearly all Japanese-made motorcycles now have their carburetors held in rubber sleeves that double as stub-manifolds, and the rest of the world's motorcycle makers follow Japan's lead when possible. With these manifold sleeves now available in a variety of sizes, there is little reason for anyone using a rigid carburetor mounting, and the equally wide availability of the Mikuni carburetor allows even less reason for employing superannuated instruments like the Amal GP. Diaphragm-type carburetors, like those universally found on kart engines, are relatively indifferent to vibration and need not be rubber-mounted, but all the rest benefit from being isolated. Even when no obvious symptoms of mixture instability appear, you may be sure that the tendency is there and of course vibration will cause a remarkably rapid wearing of throttle slides, needle and needle jet, and even the float valve. I would also unhesitatingly recommend the Mikuni carburetor as, for the moment at least, it does the best, most-delicate metering of any readily-available motorcycle carburetor. Another attraction of the Mikuni is that it is supplied in many different sizes, and backed by an excellent selection of jets, slides, needles, etc. Finally, the Mikuni is much better streamlined, internally, than it appears, and has a greater air-flow capacity, size for size, than almost any other carburetor. Consequently, there is little reason for choosing anything but the Mikuni unless you are very short of money and forced to take what you can get.

It is possible that one of the diaphragm-type "kart" carburetors may offer advantages in some specialized applications, and I know people who claim that good results are obtainable with things like the CV-series Keihin (the "constant-vacuum" carburetor found on Honda's CB350 and CB450). However, one should remember that the Keihin CV is designed to improve the broad-range performance of four-stroke engines and inherently poorly-suited to the two-stroke engine's quick-gulp intake characteristics - which means that it is something less than a perfect choice for the latter, in terms of sheer power, even though it might show some advantage on a trials engine. Similarly, the diaphragm-type carburetor was invented, and still is most widely used, to overcome the ordinary float-chamber's inability to feed fuel at steeply angled or inverted positions. There is nothing in its makeup to recommend it when you are looking for pure horsepower, and I would consider most diaphragm-type carburetors a good choice only in applications where there is so much jolting and jouncing involved that a conventional float-type instrument cannot function normally. The single exception here is the pressure-pulse carburetor developed by McCulloch, which employs the crankcase's pumping action to meter fuel-flow instead of a venturi. The pressure-pulse carburetor is thus capable of metering satisfactorily even at extremely low air-flow rates, and provides good throttle response and a broad range of power when used in relatively enormous sizes. In the original application, McCulloch's 100cc and 125cc kart engines were fitted with a pressure-pulse carburetor having a 1.375-inch throttle bore, which is much larger than would have been possible with a conventional carburetor. Subsequent development work showed that a slight constriction was necessary to create a venturi effect that would compensate for this carburetor's tendency toward high-speed lean-out, but it still is capable of combining remarkable peak power with a broad effective range. The single difficulty with pressure-pulse carburetors is that they are extremely sensitive to both cylinder displacement and crankcase compression ratios, working well only on engines for which they are specifically designed. Their air passages, which bleed crankcase pressure into and away from a chamber behind the metering diaphragm, have calibrated orifices and any variation in the conditions anticipated by the maker require a complete recalibration. It is not a job for amateurs.

Selecting carburetor throat size is enormously difficult: The four-stroke engine's carburetor may be chosen through relatively uncomplicated consideration of cylinder displacement and operating speed, but in the two-stroke engine's case there is an added difficulty introduced by the quick-gulp intake characteristic and by the overriding importance of pulsations in the intake tract. Engines with smallish intake ports and relatively long intake periods respond best to small carburetors; those having very wide, low intake ports will have shorter intake periods to provide the same specific time-area value and need a bigger carburetor-throat size if throttling is to be avoided. Frankly, unless you intend raising your engine's peaking speed very considerably, it is wise to retain the stock carburetor size even though you may want to exchange the original carburetor for a Mikuni. Switching to a large carburetor will alter tremendously the intake tract's resonant frequency and usually will require that the intake-tract length be changed to bring the sonic wave motions back into phase with the intake period. Otherwise, the loss of intake ram effect will more than offset any gains obtained through the larger carburetor's added flow capacity.

Despite all the difficulties introduced with a major change in carburetor size, the change becomes a necessity if other engine alterations work to raise peaking speeds by more than, say, 15-percent: A carburetor chosen with a power peak at 6000 rpm in mind is certain to cause some throttling when asked to deliver mixture at 8000 rpm. The problem here is, how do we determine the most suitable throat size? Everyone knows that while a bigger carburetor is sometimes indicated, "bigger" does not automatically become "better". Again, Yamaha's Naitoh and Nomura provide an answer, of sorts, with a handy formula for determining throttle bore:

$$D = K\sqrt{C + N}$$

Where D is throttle bore, in millimeters
K is a constant
C is cylinder displacement, in liters
N is rpm at peak power

For racing engines, the constant, K, is given as 0.8 to 0.9, and if you know how to work square-root problems it is easy enough to establish throat diameters for all engines. The catch here is that Yamaha's formula seems applicable only to road racing engines. For example, the Yamaha TR3 has a unit cylinder displacement of 173cc and develops maximum power at about 10,000 rpm, and if we assume that K = 0.85, then

$$D = 0.85\sqrt{0.173 \times 10{,}000}$$

$$D = 35.4 \text{ mm}$$

Thus, the calculated throttle bore diameter is only slightly larger than that actually used on the TR3 engine, which indicates that the formula probably is valid for most road racing power units. But when we apply the same formula to Yamaha's DT-1, a 250cc single developing maximum power at about 6000 rpm, the calculated throttle-bore diameter is 31mm even if we use K = 0.8. In reality, the DT-1 is fitted with a 26mm carburetor and dynamometer tests have shown that this engine responds badly to a 30mm carburetor, which means that for the Yamaha "enduro" model, the best results are obtained at K = 0.64. On the other hand, the similar DT-1MX does perform well with a 30mm carburetor, and that works out to be K = 0.69. And the Sachs 125's displacement, power peak and carburetor size gives us K = 0.91, which demonstrates that there are factors here not given proper recognition in the Naitoh-Nomura formula. I am convinced that there is a "gulp-factor", consisting of all the things that influence the shape of the intake pulse, (timing, port shape, connecting rod length, etc.), complicating this matter. Someday, I may have time to solve the problem. For the moment, I can only tell you that K = 0.8 is a safe value for road racing engines, while motocross and enduro engines may need anything from K = 0.65 to K = 0.9, with relatively low, wide intake ports (which give a brief, strong intake gulp) favoring the higher values.

A final note on carburetion: In all two-stroke engines intake pulses are very strong and the sonic wave activity considerable, which has effects both good and unpleasant. On the credit side is that the high-amplitude pulsations do make it possible to obtain very high specific power from the mechanically-simple piston-port engine by blocking blowback during the second half of the intake period. But these same pulsations also have a terrible effect on the carburetor's ability to accurately meter fuel, by leading a large part of the air drawn into the engine past the spray nozzle *three times:* Air passes the nozzle moving into the intake tract, then reverses direction as a result of the pulse generated when the intake port chops shut, and passes the spray nozzle a third time as the next intake period begins. This may sound slightly improbable -but there is absolute evidence it is happening in the fog of fuel one sees dancing in front of the two-stroke engine's carburetor. Now if this triple-passage occurred at all engine speeds, no problem would arise; unhappily, intake tract resonances - the fundamental vibration and its harmonics – slide in and out of phase with changes in engine speed. Relatively small carburetors tend to damp these resonances, and therefore are less subject to vibrations in mixture strength, which makes them particularly suitable in any application where a broad power range is more important than maximum power. This problem with mixture delivery can easily become so severe that it will be impossible to obtain clean running, especially if the carburetor is inserted at the wrong point in the overall intake tract (which consists of the port, manifold -if any - and the pipe connecting the carburetor with the air cleaner as well as the carburetor itself). Maximum air flow will be obtained with the carburetor crowded close to the port window, and an extension on the carburetor's inlet to provide the correct tract length, but that arrangement also gives the worst conditions for mixture delivery. Positioning the carburetor at the intake tract's outer end reduces volumetric efficiency somewhat, but provides the best mixture-strength stability. Connections to the air cleaner should be as short as possible, but if it is necessary to separate the carburetor and air cleaner by more than a couple of inches, the passage linking them should be either a cone (diverging at least 15-degrees) or a parallel-wall tube having about 400-percent of the throttle bore's cross-sectional area. Both of these will provide essentially the same condition as a pure, atmospheric inlet, and prevent secondary resonances that also can upset fuel metering.

IGNITION FUNDAMENTALS

Ignition systems always seem to create difficulties for those who tune high-output, two-stroke engines, and for good reason: First, the two-stroke engine's ignition system marches double-time, as compared with a four-stroke operating at the same crank speed, which not only doubles the rate of wear at the breaker points but also raises the heat input to the spark plugs. Secondly, the plugs are subjected to a fog of oil between firings that bakes into a layer of semi-conductive material over the plug insulator. Therefore, the tuner faces a situation in which the ignition system's ability to generate sparks deteriorates rapidly, while the voltage required to actually fire the plugs is on the increase. Difficulties inevitably occur, but these can be resolved - usually without resorting to draconian measures - if there is a clear understanding of ignition fundamentals.

Central to the function of all ignition systems is the high-ratio transformer we call a "coil", which comprises two sets of windings around a laminated, soft-iron core. Induction coils of this type employ the mutual influence of magnetism and electrical current: pass a current through a coil of wire and you create a magnetic field; sweep the magnetic field through a coil and you will induce a current in the wire. So in the common ignition coil, you will find primary windings to create an electromagnetic field, and secondary windings in which an electrical potential is induced. The arrangement constitutes a kind of electrical transmission, in which, for example, 12-volts are applied to about 400 turns of primary windings and amplified by the approximately-15,000 turns of wire in the secondary windings (a step-up ratio of 375:l) to the 20,000-plus volts needed to produce a spark at the plug. With 100-percent efficiency, the coil would increase the primary voltage to 45,000 volts in the secondary windings (375 x 12 = 45,000) but that is a practical impossibility. Other practical difficulties include the limitation on the amount of current the primary windings will accept without melting, and the voltage that can be. induced on the secondary side before the resistance of the internal

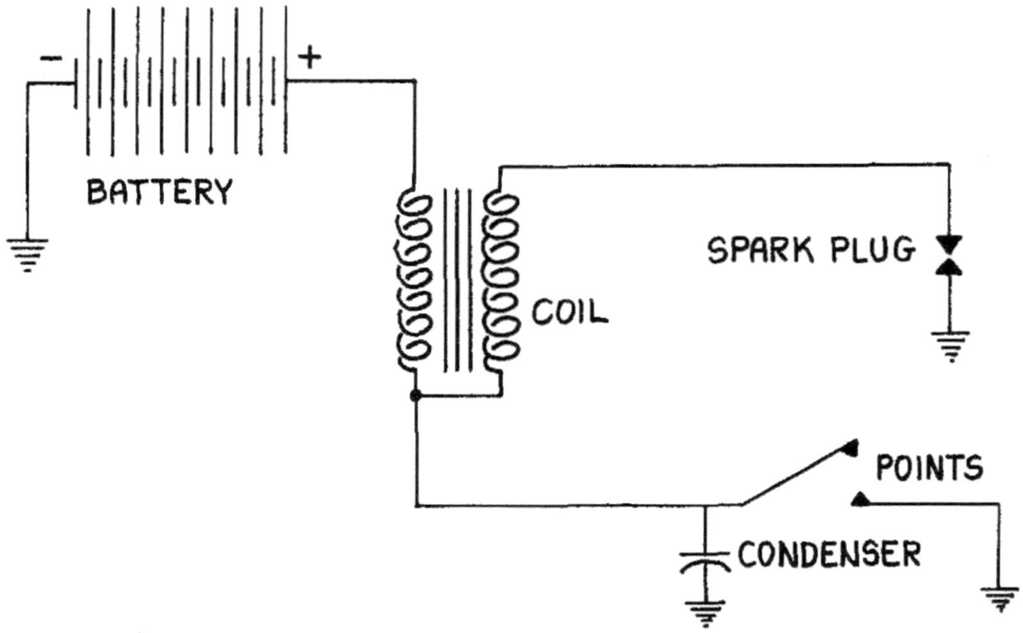

In the battery and coil ignition system, field collapse in the coil induces spark voltage when the points open. The condenser absorbs the surge from the primary windings and prevents arcing at the points.

insulation is overcome and short-circuit sparking occurs inside the coil. Also, there are limits to the electrical load the ignition system's contact-points will carry, and extremely high secondary voltages can cause rapid electrical erosion of the spark plug's electrode.

All of the presently-used ignition systems feature the high-ratio transformer, or coil, just described; there are several ways of triggering that coil into action and these can have a considerable effect on ignition performance. The system most commonly employed on touring engines is one in which current is drawn from a battery, passes through the primary windings, and then through an engine-driven switch (the "points") to ground –which completes the circuit, as the battery has one terminal connected to the motorcycle's frame, or ground. It should be understood that the ignition points are normally closed, and that a spark is obtained when the points are bumped open by the contact-breaker cam. There is a voltage rise in the secondary windings when the points close and current starts to flow through the primary side of the coil, but due to a phenomenon called, "impedance", the magnetic field builds too slowly to induce maximum voltage. Only when the points open, and the primary current disappears, does the field move quickly enough - in collapsing - to induce ignition voltage. Because there are so few primary-winding turns, compared with the secondaries, the field collapse creates a very much lower voltage there, but it would be high enough to cause an electrical arc to form between the point contacts, which have only barely opened, except for the action of the condenser. This last item acts as a kind of electrical reservoir, and it receives the induced primary current and prevents voltage from rising high enough to form an arc across the opening points. Should the condenser fail, an arc will be established, and the effect is rather like having the points closed, with the current in the primary windings reaching a level sufficient to choke back the magnetic field's collapse and thus drastically reduce the voltage in the secondary side of the coil.

Viewed strictly as regards its spark-producing ability, the old battery-and-coil ignition system is limited, but is not nearly as bad as present-day enthusiasm for the more elaborate alternatives may make it appear. It requires that you carry a battery on the motorcycle, and that battery must be rubber-mounted to prevent vibration from shaking its plates apart, but a battery large enough to energize the ignition system for the duration of a race is little heavier than a magneto and, moreover, does not borrow from the engine's power output. You must, of course, keep the battery on a low-rate charger between race meetings to prevent sulfation of its plates, and if you are the forgetful type that may be counted a short-coming. Still, the battery-and-coil system's single serious deficiency is simply that it does not produce sparks quickly enough. By that, I do not mean that it is slow in terms of the total number of sparks it will yield in a given time period (it is capable of making at least 10,000 sparks per minute), but simply that the voltage to produce a single spark builds rather slowly. That characteristic is most unfortunate when combined with the perpetually semi-fouled condition of the two-stroke engine's spark plugs, for if spark voltage rises too slowly, the electrical potential in the coil's secondary windings will leak away across the deposits on the plug insulator and voltage will not climb to the level required to ionize the gases in the gap between the electrode and the ground wire. In that instance, no spark occurs, even though the basic system may be capable of delivering voltages of 30kv (30 kilovolts, or 30,000-volts), which certainly would be enough to produce ignition with a clean plug. Indeed, touring engines running on relatively lean mixtures, with only minimal amounts of oil passing through, and with "hot" plugs perform very satisfactorily when sparked by the humble battery-and-coil system. Racing engines, however, must be fed rich mixtures, lots of oil, and be fitted with decidedly cold plugs – which means that they often will run without the slightest sign of distress on wide-open throttle (a condition that keeps the plugs burned clean) being supplied sparks from a battery-and-coil system, but lapse into a fit of misfiring the first time the throttle is closed.

A magneto is the most commonly-applied alternative to the battery-and-coil ignition system. Many people assume that the single virtue of the magneto is that it

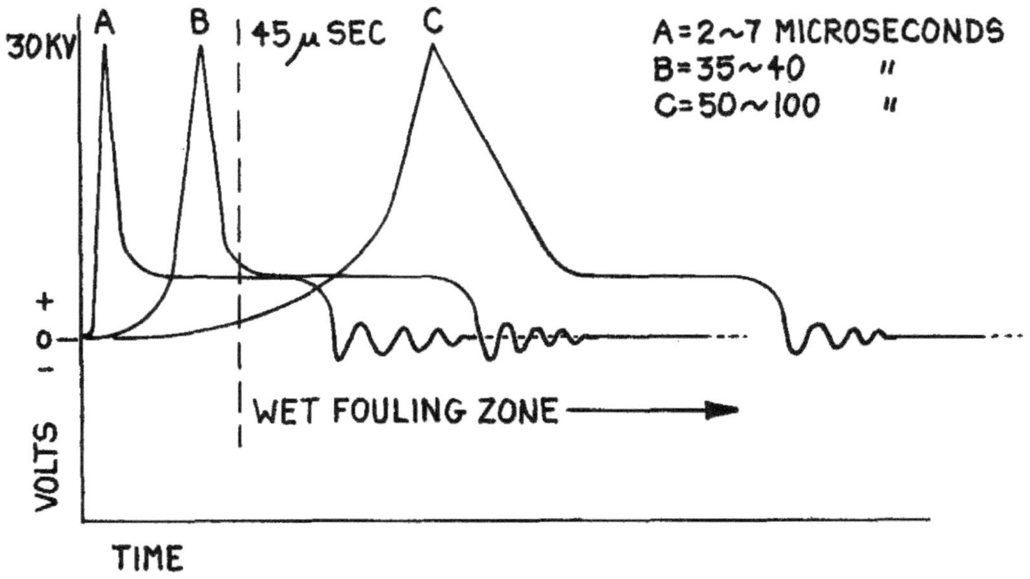

Capacitor-discharge ignitions deliver spark voltage too quickly for wet fouling to occur (A) as does a good magneto (B) but the voltage rise-time for battery/coil ignition is too long (C) and fouling will occur even though its peak potential is equal to the others.

eliminates any need for carrying a battery on the motorcycle, but in the context of racing that aspect is merely a minor convenience. A far more important justification for magneto ignition is that it provides a far more rapid voltage rise than a battery-and-coil system. Racing plugs have a very short insulator nose, which means that they also have poor resistance to "wet" fouling- the kind of spark-voltage short-circuiting that occurs when the plug nose is coated with soot and raw gasoline. This mixture isn't what you'd call a good conductor of electricity, but it is good enough to drain off the slow-rise voltage delivered by battery-and-coil ignition system. The critical voltage rise time seems to be about 45 microseconds: if the electrical potential across the spark gap reaches 15kv in 40 microseconds or less, misfires due to wet fouling rarely occur. And a good magneto delivers maximum voltage to the plug (a maximum that may easily exceed 30kv) in only 35-40 microseconds after the system is triggered.

The explanation for this faster rise-rate is found in the manner in which magnetos make sparks: Energy is supplied by an engine-driven, rotating magnet, which sweeps its field through a coil, or coils, to generate the operating voltage. In some magnetos all of the coils are contained within one housing, but most motorcycles have "energy transfer" magnetos (originally developed for high-altitude aircraft engines). These have low-tension generating coils connected in parallel with the breaker points and the spark coil. The points, and the phasing of the rotating magnets relative to the generating coils, are timed together, for the points must open when the magnetic flux is strongest. With the points closed, and the magneto rotor swinging into position next to the generating coil, the rotor's magnetic field sweeps through the coil and induces a strong current. This current is, in effect, short-circuited through the points, so that the wire in the coil passes the electricity around in a continuous loop, and that flow of current makes the coil into an electromagnet with a field polarity opposite that of the rotor. So you have two magnetic fields pushing against each other, and that condition persists until the points open. Then the current in the generating coil is interrupted, and its magnetic field collapses, which permits the rotor's field - at that moment forced back like a spring - to snap through the coil. The extremely rapid flux change induces an extraordinarily strong voltage surge in the generating coil, and this is communicated to the primary windings of the spark coil-which responds by producing enough voltage to fire the spark plug. Substantially the same process occurs in the traditional magneto, in which the generating coil is also the spark coil's primary winding, and is overlayed by the high-tension windings.

The principal reasons for using the E.T. magneto are that A) the spark coil is removed from direct proximity with the engine, where it frequently becomes too hot for its own survival, and B) it allows you to keep the length of the high-tension spark lead to a minimum. This last is especially important, for while electrical resistance is

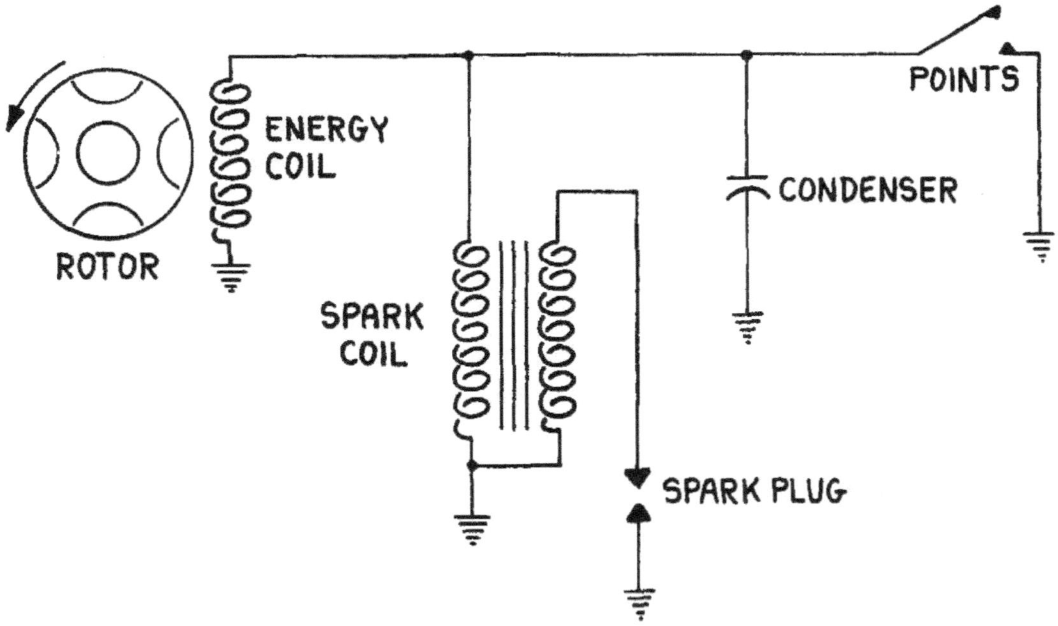

Energy transfer magnetos use a magnetized rotor and energizing coil to hit the spark coil with a voltage surge when the points open. Obviously, the energizing pulses must be timed with the points.

unimportant here, ignition leads have capacitance and inductance properties that adversely effect voltage delivered to the spark plug and should therefore be kept as short as possible. By and large, magnetos arc very reliable, but when trouble occurs, look for either pitted (or dirty) points, a faulty condenser, or - more rarely - a weak coil. Especially, make every effort to keep the points clean, smooth, and adjusted. The contact surfaces have to carry a considerable electrical load, and if they become burned, or filmed with oil, spark voltage drops very sharply. After all, the voltage induced in the spark coil is in the final analysis a product of the flux change down at the field coil, which is in turn largely a function of the current strength in the field coil just prior to the points' opening. Oil films and pitting increase the resistance across the point surfaces and reduce the current strength. Condensers fail mostly because the condenser usually is tucked away in a housing on the engine next to the points. There, they become overheated, their internal insulation softens, and short-circuiting follows. In my opinion, a racing engine's condenser (s) should always be mounted externally, even though that arrangement may not be visually tidy. I prefer placing the condensers next to the, spark coils. All of the magneto's elements are connected as a parallel circuit, so you can simply attach the condenser leads to the spark coil's input terminal. I might mention here, too, that it is unlikely' that you will find a condenser too large to give good results on the magneto. I have used the big Mallory condensers in many applications and find that there often is an improvement in ignition performance with the added capacitance. Why? Probably because the larger capacity reduces even further the slight arcing at the points, and gives a cleaner termination of the field coil current.

Sheer mismanagement probably is the most potent source of all problems encountered with magnetos. As noted earlier, the magneto's performance is strongly dependent on having the points break right at the moment of maximum flux, which means that when you are timing the points to deliver a spark at the correct moment in terms of crankshaft rotation, you must also "time" the magneto's rotor and stator. Usually there will be an indexing mark on the rotor to be aligned with a similar mark on the stator; otherwise you will need a manufacturer's manual dealing with the magneto's construction to give you instructions as regards the orientation of rotor and stator. You can adapt a magneto from one engine to work on another, but the conversion isn't always easy: One difficulty arises because the air-gap between the rotor and stator (the stator is the laminated core for the generating coil) is very small, and while the rotor will center more or less automatically on the crankshaft unless the remachining of tapers is incredibly sloppy, getting the stator mounting concentric with the rotor can be quite a job. Whatever else you do, just don't compensate for misalignment by trimming away at the stator's poles, as a very small increase in the air gap will cause a marked sagging of the magneto's output. In fact, you shouldn't even drill the stator's mounting holes oversized when making the installation. As it happens, the magnetic field shifts that induce an electrical current in the magneto's windings do the same job in the stator, and there is the same creation of an opposing magnetic field, only this one cannot be controlled by the contact breaker points. For that reason the stator - which could more conveniently be made in one piece - is assembled from a number of thin laminations. With the induced current divided and confined, the magnetic effects are minimized, but if you drill through the stator laminations, or grind away at its pole surfaces, you break through the coatings of varnish that insulate between the laminations, establish contact, and convert the stator into another electromagnet. Of course, the spark coil laminations are similarly insulated, for similar reasons, and cutting through them also is poor practice. These things are commonly done, out of ignorance of the consequences -which usually are not so severe as to be responsible for outright misfiring but should be avoided in the interest of getting the best performance possible from any given ignition system. Finally, I would caution against simply assuming that bolting a coil to a motorcycle's frame automatically insures a proper ground. On many occasions I have been asked to help solve a persistent and mysterious ignition problem, only to find that the single difficulty was an incomplete circuit: the coil had been bolted to the frame with a fine disregard for the insulating properties of paint. Persistent coil failure is another complaint frequently heard, particularly with reference to the old internal-coil magneto. Overheating of the coil usually is at fault here, but you should be aware that those who made the coil planned for the temperatures involved and internal short-circuiting usually will not develop unless there is an electrical overload. The overload comes when a spark lead is lost, and the voltage from the coil no longer has a path to ground via the spark plug. If conditions are such that the electrical potential between the plug's electrode and ground wire ionizes the gases there when coil voltage is at 15kv, then that is as high as the voltage will go, because all the coil's energy will discharge through the plug and most of the discharge period will be at a voltage very much lower than that present ns the spark was formed. But if the spark lead is disconnected for any reason, the coil's internal voltage, will climb extremely high, and may become high enough to force through the insulation. When that occurs, a permanent path to ground is established and much of every successive voltage surge will burn its way across the same path. The result is a sharp loss of voltage at the plug, and the urgent need for a replacement coil. Heat will soften the insulation inside the coil and make short-circuiting easier, but more often than not the trouble would have been avoided had the tuner been careful to keep the leads in place - and to ground them in some manner when motoring the engine (by pushing the bike) with the spark plugs removed.

"Transistorized" ignitions are becoming more popular, and in some circles are regarded as the sovereign remedy for all ignition ailments. They are something other than their reputation suggests. Ordinary battery-and-coil systems can be transistorized, by using the points to valve "trigger" current to a transistor - which actually carries the electrical load. This kind of system offers no particular advantage for most motorcycle ignition systems, as its principal virtue is that it will handle the very high primary currents encountered in multi-cylinder automobile engines. Also, the points will last almost indefinitely, as they are under only a micro-amp load. The basic transistorized ignition system can be embellished to include a magnetic triggering device that replaces the mechanical contact breaker points entirely, and the advantage here is that all problems with point-float are eliminated and the sparks may be delivered with great accuracy. A disadvantage is that the circuitry becomes rather complicated, and an external power source (a battery) is required. The transistorized battery-and-coil ignition system with magnetic triggering is capable of delivering an extraordinary number of sparks per minute, but the character of individual sparks remains the same as is obtained with conventional points and I am not persuaded that this type system offers any advantages in terms of high-speed, high-output two-stroke engines.

There are advantages with other types of transistorized ignitions, if other of the system's elements are properly arranged. As in the case of the capacitor-discharge system, which can be made to be battery operated, or as a magneto-type device. Either way, there is a fairly conventional spark coil, and the spark is produced by the discharge of the capacitor's electrical contents into the coil's primary windings. Usually, the capacitor will be charged to about 400 volts, which is enough to produce a. more-than-adequate voltage at the coil's output lead. But the capacitor-discharge system's best feature is that it is tremendously fast, faster than even the conventional magneto: Something in the order of 50-100 microseconds are required for a single electrical pulse from a conventional battery-and-coil to reach sparking strength; a magneto requires 35-40 microseconds (which is just below the wet-fouling level of 45 microseconds); but a good capacitor-discharge system takes only a couple of microseconds in climbing to full voltage, which means that it is so far below the wet-fouling level that it may be considered virtually immune to that disorder. Moreover, its spark is delivered so rapidly that firing will occur even when the plug is very badly fouled with oil or anything else. The single disadvantage of the capacitor-discharge ignition system is that it can deliver this terrific spark so briefly that ignition of the air / fuel mixture in the combustion chamber does not follow. Magnetos, for instance, keep their sparks going for about 0.5-1.0 millisecond, while the capacitor-discharge system's spark may last less than a tenth of that time. It has become common practice to add a resistor in the circuit to prevent the capacitor from dumping too fast and delivering a too-brief spark to the plug, but there still are systems being made without this feature - which is to be regretted, as cold-start problems will occur even with everything else in order, including a fresh, clean plug. A too-brief spark is also ,likely to be unable to fire the kind of rich mixtures needed by two-stroke racing engines.

Sports / touring motorcycles, like the 750cc, 3-cylinder Kawasaki usually have a battery powering their capacitor-discharge ignition system. The battery's 12-volt potential is multiplied to the required 400-volts by a special oscillating circuit and transformer, and the capacitors dumped into the spark coils via transistors of the type called SCR (silicon controlled rectifiers). The SCRs are actuated by separate circuits, which are in turn controlled by magnetic trigger. Conventional points will also work here, but most manufacturers seem to feel that as long as they have to resort to solid-state electronics, they might as well go all the way and dispense with the breaker-points. A more suitable system, for racing, is one in which the capacitor (or condenser) is charged directly from a magneto-type rotor/stator arrangement with an appropriate generating coil. By inserting a rectifier-type transistor between the generating coil and the capacitor, voltage is trapped in the latter, to be used in activating the spark coil. With a little jiggling of circuitry, it is possible to use the same rotating magnets to charge the capacitor, and trigger an SCR into releasing the charge to the coil, and this is the arrangement you would find inside the encapsulated wiring of most current "CD magnetos" if you could get them apart. There isn't much I can tell you about these ignition systems except that when they work they work very well, and when they stop working you are obliged to start replacing magic-boxes because everything in their inner workings is A) sealed off so you can't touch the transistors, resistors, etc., and B) you have to know more about electronics to fix the things than most people have at their command. I have done some work in this area, but I wouldn't feel comfortable trying to instruct you in the intricacies of solid-state electronics even if that did not require more space than can be given the subject here - my ignorance is too great. One thing I can tell you is that all the magnetically-triggered systems should be timed using a "strobe" light; it is impossible to set them accurately by any other means, and the best of these systems, badly timed, is inferior to the conventional battery-and-coil, or magneto. Proper timing is the first requirement with *any* ignition system.

Finding the exact ignition advance yielding best results with a given engine is, as noted at the beginning of this chapter, a time-consuming chore, but one worth pursuing diligently. Actually, unless you have changed your engine very considerably from the stock specifications you probably will find that maximum performance is obtained at a setting very near that recommended by the engine's manufacturer. Generally speaking, modifications that tend to raise engine output without increasing the crank speed at which maximum power occurs will require that the spark be retarded slightly from the standard setting. Major upward relocations of the power peak usually require a more advanced spark. I would be delighted to provide you with a universally-applicable rule for predicting the exact spark timing for two stroke engines; unfortunately no such rule exists. It is known that about 75-percent of the combustion process should be completed by the time the piston has reached TDC, and that when the average rate of flame-front travel can be found, calculating spark advance from the flame propagation rate and the distance from the spark plug to the remote end of the combustion chamber is a simple arithmetical problem. But confusion arises because flame propagation rates vary so enormously. Both the type and location of the spark plug has an influence in this, as do pre-combustion turbulence and the shape of the combustion chamber. Engine speed also has an effect, and as your modified engine probably departs from the original specifications in all of these particulars, it is essential that you do the careful testing necessary in finding that elusive optimum. Start testing with the ignition timing retarded about 5-degrees from the stock setting, and then advance the spark in two-degree increments until best results are obtained. Dynamometer testing is the best means of verifying results, but you can also use a drag strip, for it has been found that while sheer rider-technique is the determining factor in elapsed time, the motorcycle's speed at the end of the quarter-mile is almost exclusively a function of engine horsepower. Thus, if you find that maximum speed (which may be taken from an accurate tachometer) is reached with a spark advance of, say, 27-degrees BTC, then that timing will be the one providing maximum horsepower. Interestingly, the influence of mixture turbulence - which increases with engine speed-is so strong that the spark-advance optimum does not change much even over a fairly wide speed range. Indeed, from all available evidence, the ability of combustion-chamber turbulence to speed burning is so great that increases in peaking speeds frequently require a slight retarding of ignition, while advancing the spark may well give a small improvement in power at lower-than-peak engine speeds at the expense of maximum power.

SPARK PLUGS

People who are relatively inexperienced in the art of tuning racing two-stroke engines seem always to fall into the trap of trying to use a spark plug that is too "cold", and / or one having the wrong nose configuration, which they then compensate with a too-lean mixture. This kind of error probably stems from a basic misunderstanding of spark plug heat-range, and the reason for having more than one type of plug. The reason?

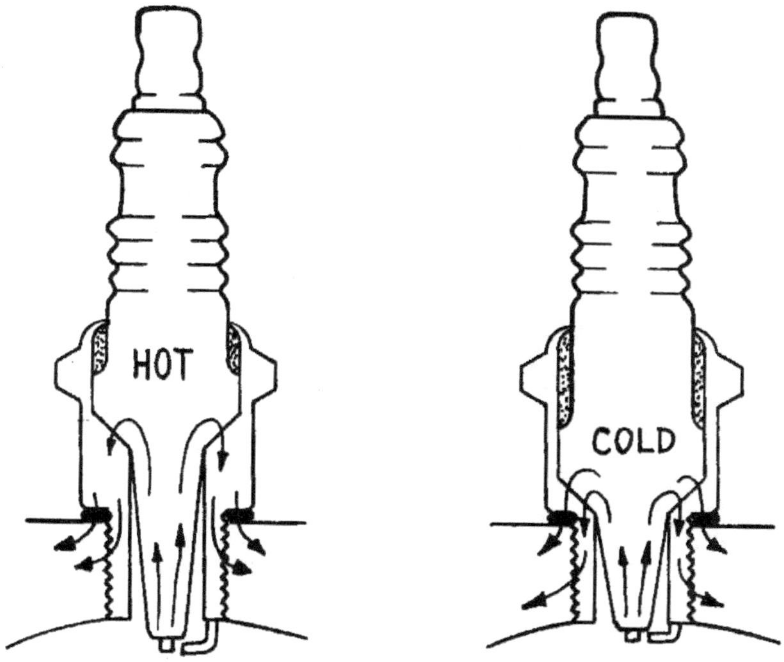

Spark plugs look the same externally, but the heat path from the insulator tip through the plug body into the cylinderhead is much longer in the "hot" plug than in the colder ranges.

As it happens, the temperature of the burning mixture in the combustion chamber is high enough to melt the engine, and it fails to do so only because heat is carried away from the combustion chamber's walls fast enough to prevent them from reaching flame temperature. The spark plug reaches much higher temperatures than the cylinder head itself, because heat moving away from the plug's nose must cross the joint formed, by the threads on the plug and in the plug hole. And, of course, the exposed plug electrodes are separated from the cylinder head's cooling fins by -a very long heat-path, and in the case of the center electrode, by the plug's insulator. Consequently, the electrodes get extremely hot, and that is very much a mixed blessing: Fairly high temperatures are required to keep the plug's electrodes and insulator tip burned clean, to prevent the formation of sooty deposits that can short-circuit the spark. But if the electrodes are allowed to become too hot, they will constitute an independent and uncontrolled ignition source. That is to say, the electrodes may become hot enough to ignite the air/fuel mixture in advance of the spark. When that occurs, combustion chamber temperatures climb even higher which causes auto-ignition still earlier and that yields yet higher temperatures in an ascending spiral that quickly leads to outright engine failure.

So, the correct spark plug for a given engine is one that stays hot enough to burn its electrodes and insulator tip clean, but does not reach temperatures high enough to cause auto-ignition (more commonly termed, "pre-ignition"). Unfortunately, all spark plugs transfer heat from their electrodes to the cylinder head into which they are threaded at a rate fixed by their construction, while engine operating temperatures vary enormously. "Correct" selection of a spark plug is therefore a matter of choosing one having a heat-transfer rate compatible with the application at hand. The process of selection is less exacting today than in the early days of the internal combustion engine, principally because the mica and porcelain used as insulators have been replaced by aluminum oxide ceramics that are stronger and much better conductors of heat. Other improvements in spark plug construction have further widened plug heat-range, hut not to such extent that any plug will perform satisfactorily in any engine. Touring plugs have extended tips, which burn clean but have exposed ground electrodes and center electrodes, and these respond to prolonged full-throttle operation by becoming incandescent; racing plugs have short, shielded electrodes and will tolerate higher combustion chamber temperatures but foul very quickly under starting and idling conditions.

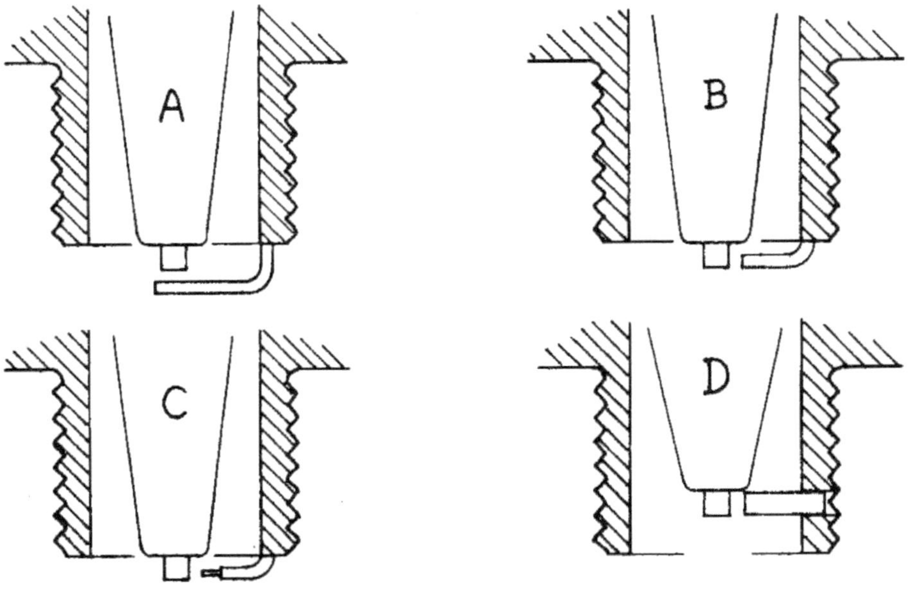

Spark plugs commonly have the ground wire overlapping the center electrode (A) but in plugs made for two-stroke engines it approaches from the side (B). A variation of plug "B" has a short platinum wire in the end of the ground (C). Recessed-nose, side-wire racing plugs (D) are prone to fouling and should be used only when an extremely cold spark plug is an absolute necessity.

The most common spark plug configuration is the one in which the ground electrode extends over the end of the center electrode. A much better type, in any two-stroke engine application, has the ground electrode reaching in slightly lower to approach the center electrode from its side. The improvement, here, is that the ground electrode is somewhat shorter, and being shorter offers a more direct heat path to the plug body - which means that it is less likely to become white-hot. A refinement of this design has a short bit of platinum wire inserted in the tip of the ground electrode. This wire's diameter is quite small, and it is used in conjunction with a center electrode also reduced in diameter at its tip, an apparently minor difference unless you consider that much less voltage is required to form a spark between two points than between two flat surfaces. An ordinary iron ground electrode cannot be made pointed, because its tip would instantly overheat and melt, but platinum -with a melting point of 1774°C, as compared with 1535°C for iron - is less likely to melt, has a thermal conductivity 18 times greater than iron, and will not oxidize. As a result, the platinum electrode survives its hostile environment very well even when used in small diameters, and I would not hesitate to recommend the platinum-tip plug for most racing applications. At any given heat-range, the platinum plug will provide longer life and less tendency to foul than any conventional type plug. Pure racing plugs, which have deeply recessed insulators and center electrodes, with an extremely short ground electrode bridging straight across from a hole through the side of the plug body, should be used only as a last-resort in two-stroke engines. Racing plugs of the type described are very, very prone to wet fouling, oil fouling and every other kind of fouling possible. They are a necessary evil in hyper-horsepower road-racing engines, but should never be used where extended-nose plugs will serve. Not unless there is some kind of major breakthrough in either spark plug or ignition system design, which always is a possibility. I would recommend that all who are serious about keeping ahead of the pack keep abreast of developments in these areas. All the spark plug manufacturers are working constantly to give us improved foul resistance and reduced spark-voltage requirements, and they are very accommodating to anyone who takes the trouble to write and ask for literature.

"Reading" spark plugs, and the process of selecting correct heat-range, falls much more into the realm of art than science, and it is an art in which one becomes really proficient only after long and sometimes painful experience. But there are a few rules that may be used for guidance by those who have yet to acquire experience - or by the many whose experience has left their ignorance largely undiluted: First, you should know that it is all but impossible to read anything in the appearance of a spark plug unless the engine has been cut clean after having been brought up to operating temperature and given a long burst of wide-open throttle. Very experienced tuners will see the signs they're looking for under the layer of soot, oil and fuel that 'accumulates so quickly at idle, but even they vastly prefer to work with clean-cut plugs. Second. get the right heat-range before you try to rend mixture strength, and my recommendation is that you always use the hottest plug the engine will tolerate. You'll know a plug is too hot when you observe signs of blistering around the insulator nose (which will also be scorched white) and on the electrodes. A too-lean mixture will also give you a whitish insulator, but will not usually produce the burned, pitted appearance of the electrodes that is characteristic of a too-hot spark plug. Also look for signs of melting along the sharp edges at the ground electrode's end – any sharp corner will get hotter than other areas along the electrode, and trouble will first be revealed there. A plug that is too cold simply looks, and is, wet. Plugs of the correct heat-range get hot enough to burn away oil, and soot, and will have only dry, brown to tan deposits on their insulators after a hard run. As noted before, the correct mixture strength will be wry slightly leaner than that which is just lean enough to keep the engine from four-stroking. How much leaner? Not very much, and until you have gained considerable experience with a particular engine you should not reduce strength below the jetting that provides clean running. To get a bit closer to the optimum, I watch the faint, almost invisible ring of soot that forms around the electrode on the insulator's nose, and the light dusting of soot over the exposed end of the plug body. There is a point at which I see "just enough" soot, and if there is more or less than that - I interpret the signs to mean a mixture that is too rich or too lean, respectively. And I couldn't begin to tell you how much soot is "just enough"; that would be like trying to explain a taste, or sound, or smell. With experience, you learn to recognize what it is you're looking for, and there is no substitute for that experience. You will also learn - if you know where to look - that the faint light spot on the electrodes at the sides of the spark gap are an important clue to ignition system performance. When you have a spot of about the same diameter as the ground electrode's end showing on the center electrode, or *vice versa,* then you may be sure the magneto is doing its job. When that spot begins to fade, or become ragged around the edges and shrinking in diameter, the ignition system isn't performing as it should. Finally, with experience you'll learn to give your very close attention to *all* aspects of the mundane task of selecting jets and plugs, and spark timing, because in these things you ultimately succeed or fail as a tuner; all the rest is mere mathematics, surgery and wrench-twirling.

Two Stroke TUNER'S HANDBOOK

Notes

Notes